D0699809

WHAT MADE GERTIE GALLOP?

WHAT MADE GERTIE GALLOP?

Lessons from Project Failures

O.P. KHARBANDA, D.Che.
JEFFREY K. PINTO, Ph.D.

VNR VAN NOSTRAND REINHOLD
I(T)P ™ A Division of International Thomson Publishing Inc.

New York • Albany • Bonn • Boston • Detroit • London • Madrid • Melbourne
Mexico City • Paris • San Francisco • Singapore • Tokyo • Toronto

I(T)P™ Van Nostrand Reinhold is a division of International Thomson Publishing Inc.
The ITP logo is a trademark under license.

Printed in the United States of America.

For more information, contact:

Van Nostrand Reinhold
115 Fifth Avenue
New York, NY 10003

Chapman & Hall GmbH
Pappelallee 3
69469 Weinheim
Germany

Chapman & Hall
2-6 Boundary Row
London
SE1 8HN
United Kingdom

International Thomson Publishing Asia
221 Henderson Road #05-10
Henderson Building
Singapore 0315

Thomas Nelson Australia
102 Dodds Street
South Melbourne, 3205
Victoria, Australia

International Thomson Publishing Japan
Hirakawacho Kyowa Building, 3F
2-2-1 Hirakawacho
Chiyoda-ku, 102 Tokyo
Japan

Nelson Canada
1120 Birchmount Road
Scarborough, Ontario
Canada M1K 5G4

International Thomson Editores
Seneca 53
Col. Polanco
11560 Mexico D.F. Mexico

1 2 3 4 5 6 7 8 9 10 QEBFF 02 01 00 99 98 97 96

Library of Congress Cataloging-in-Publication Data
Kharbanda, Om Prakash.
Pinto, Jeffrey K.
What made Gertie gallop? : lessons from project failures / O.P. Kharbanda, Jeffrey K. Pinto.
p. cm.
Includes index.
ISBN 0-442-02158-5
1. Industrial project management. 2. Industrial project management—Case studies. 3. Industrial project management—Developing countries—Case studies.
I. Pinto, Jeffrey K. II. Title.
HD69.P75K553 1996
658.4'04-dc20 96-11873
CIP

Dedication

Lovingly dedicated to Sakshi (means "witness") who came into the world (March 6, 1995) just in time to "witness" the contract for this book.

and

For Emily and AJ, two ongoing "projects" for whom love is the best safeguard against failure.

Acknowledgments

To our friends worldwide who have helped us in many ways. Most must remain anonymous either by their choice or due to lack of space. But a few MUST be named.

First and foremost are our life partners, Sudershan and Mary Beth, usually 'silent' in our professional work, though Sudershan is now a co-author and a co-faculty in her own right. She even 'teaches' English to her husband and helps 'unsplit' many split infinitives!

Our present effort was accomplished entirely through e-mail in record time. This follows the same course that we took in our first endeavor, Successful Project Managers: Leading Your Team to Success, VNR, 1995. For this, we owe a special debt to Dr. S. Ramani, Director of the National Centre For Software Technology, India, and a prime mover of the educational network, ERNET—India's window to the Internet.

It is to our amazement and sheer wonder how our drafts

moved back and forth half-way around the world from our respective work stations—duly edited and transposed (as necessary)—and in due course (sooner than was ever ossible) finalised. It seemed to be far more reliable than the good old fashioned snail-mail, provided you don't forget your password! The future in this aspect promises to be even more exciting! Readers may be interested to know that we have never met, though we 'meet' almost every day/night!

One of us, having been born in the 'low-tech' era, finds the high-tech too high for him. But thanks to the instant, every-ready help of his son, Vivek, a CPA, who owns and operates a computer centre with American collaboration, he finds himself getting more familiar with this new, high-tech world. The 'fat' manuals just scare the father, but Vivek can browse through it all in minutes and tell his Papa what key to press, and when!

Last but not least our publisher's team comprising of John Boyd (Editor and Publisher) and Jacqueline Jeng (Marketing Manager) and their associates. They have seen this project through, from inception to completion. Even the title bears their stamp of approval. They have helped translate our 'labour of love' to a magnificient reality that you now hold in your hands!

We are looking forward to hearing from YOU about this book and any related matter. We promise to answer any of your queries, IF we know the answer! We may even use your material in the form of case studies in our future writings—with prior permission, of course.

Om P. Kharbanda, Mumbai, India. opk@soochak.ncst.ernet.in
Jeffrey K. Pinto, Erie, Penn, USA. jkp4@psuvm.psu.edu

Preface

On apprend en faillant—One learns by failing.
French Proverb

When the Tacoma Narrows Bridge (known as "Galloping Gertie" for the strange, aerodynamic gyrations it emitted during its short life) collapsed in the fall of 1940, it did not mark the beginning of any special, new era of project failures. Projects, including bridges, have been failing since the dawn of time. Rather, the "magnificent" disaster embodied in the collapse of the Tacoma Narrows Bridge serves as an abiding testimony to both humanity's quest for perfection through technology and the knowledge that such a quest can only proceed on the back of failure. Much as a child learns to walk through trial and error, standing and falling relentlessly until the process becomes ingrained, so too does technology typically advance through the knowledge gained from disaster.

In the continuing quest for better project management skills and techniques, experience plays a crucial role. It is, however, a two-edged sword, consisting of both positive and negative consequences. And yet, even our failures have their use. We gain wisdom every bit as much from failure as from success; we often discover what works by finding out what does not. Indeed, it can reasonably be argued that those who never made a mistake also never made a discovery. Horne Tooke used to say of his studies in intellectual philosophy, that he had become all the better acquainted with the country through having had the good luck to lose his way. In a similar vein, many of us count as our most valuable management lessons those that were learned as the result of failure.

Why study project failures and disasters? There are specific lessons to be learned, of course, as such studies also yield valuable data in relation to future projects. Unfortunately, the cost of these mistakes is usually painfully high. Real life case studies of project disasters can be an invaluable source of information and provide a real insight into the way in which mismanagement can totally negate an otherwise successful project undertaking. One has only to make a cursory search of current management literature to see compelling examples of projects that have failed, usually with serious consequences for their organizations. For example, Borland's upgrade of DBaseIV was so poorly managed some years ago that the product had to be removed from store shelves for comprehensive debugging, costing the company tremendous expense and customer goodwill.

Much has been written about success in management, and the latest edition of the annual *Cumulative Book Index,* which chronicles all books in print published in the U.S. for 1993–1994, lists over a hundred titles relating to success and associated topics, but less than ten addressing the subject of failure. Such is the lack of coverage of failure, the subject we have singled out here for study in the hope that we may learn to manage projects better. The clash of giants such as Lockheed and Rolls Royce in 1971 resulted from a more than ten-fold cost overrun on a single project. Currently, Denver's new multibillion dollar international airport is plagued by so many technical problems that its opening date has been repeatedly pushed off, costing the city and airport authority over $1 million a day in late penalties and interest.

Such overruns in time and cost are customarily seen as a failure in project management and when extreme can well be a "disaster." Unfortunately, in many projects, particularly large-scale projects, they are the norm rather than the exception that they ought to be. These operational errors are frequently compounded when, rather than learning from past mistakes, project disasters are quietly filed away rather than studied. Hence, our particular interest in project failures: instead of simply telling readers "how to do it," we show "how not to do it!" Our analysis of specific projects should enable the reader to learn from the blunders of others— by far the cheapest way to learn.

We have seen, in our research and consulting experience, that most companies spend thousands of hours to plan and implement a multimillion or even multibillion dollar investment, but far too little time to critically evaluate and learn from their experiences, e.g., asking simple but vital questions:

Was the investment worthwhile?
Did it go according to plan?
If yes, how and if not, why not?

Each project is, to some degree, unique and each therefore differs greatly from traditional, functional business tasks which are usually repetitive. As a result, each project, if poorly managed, can have an immediate, negative effect on the parent company. This book, *What Made Gertie Gallop? Lessons from Project Failures,* is designed to fill an important and heretofore, previously unexploited niche in the manager's library by offering a number of in-depth cases on well-known or large-scale projects that failed through mismanagement. We intend to offer a practical guide in project disaster and the managerial lessons that can be taken from these failures.

Learning is not easy—for people any more than for organizations. Neither of them seem to have much of a "memory" if we are to judge by the constant repetition of the same types of mistakes leading to similar disasters. Indeed, three of the worst tragedies ever, Bhopal, Exxon Valdez, and Challenger are constant reminders of this fact. In all cases, the agencies concerned were utterly confused just after the event, and the numerous contradictory statements made the confusion even worse. We venture to suggest that the real causes in these and other similar cases may never be known, thanks to human ingenuity to sweep facts under the carpet. This is indeed a pity. Headlines in the international press excoriated the organizations concerned (Union Carbide, Exxon, and NASA) as singularly ill-prepared to cope with such disaster. Hopefully our present effort across half a world away will help shed greater light on a significant problem, coping with, anticipating, and correcting project disasters before they pass the point of no return.

Approach

Our approach will be a combination of theory and case study. In the first section of the book, we will seek to lay a groundwork foundation using some important guiding principles from the field of project management to put the idea of project failure within its proper context. For example, it is necessary to first define what is meant by a disastrous project in terms of factors such as spiraling costs, sliding schedule performance failure, and lack of customer acceptance. The reader needs to be well grounded in some basic theory of the goals of projects and the constraints that managers face, particularly in (but not limited to) the case of large-scale mega projects. Once the reader understands the yardstick by which projects must be measured as well as some of the key principles for project success, we can begin examining specific examples of project failure. Each of these failures will be examples of projects that failed through mismanagement or misjudgments of a profound nature. It is not terribly useful for the reader to study projects that failed through "acts of God" or other reasons beyond the project manager's control. Finally, in the last section of the book, we will examine some general conclusions that can be drawn from the study of project failure. What are the management lessons to be learned? How can I apply these lessons to my own work situation to become a better project manager? These and other questions will be addressed in our conclusions.

In a nutshell, our intention is to help project managers do a better job of running their projects through the valuable lessons that can be learned from observing and studying failure. As someone once noted, "Everyone learns from mistakes, but nobody ever said they had to be *your* mistakes."

Table of Contents

PART 3 THE DEVELOPED WORLD IS NOT MUCH BETTER

PART 4 LET'S LISTEN AND LEARN

Part 1

Error on Trial—Setting the Scene

1

Failure as a Step to Success

INTRODUCTION

We learn from success, but we can learn much more from failure. Further, it is far better, cheaper, and of course, wiser to learn from other people's failures rather than our own. Making this point in a formal manner, the great German statesman, Otto von Bismarck, observed: "Fools you are . . . to say you learn by your experience . . . I prefer to profit by others' mistakes and avoid the price of my own."

Though there are benefits to be had from learning many business lessons the hard way, it is not an argument for making mistakes easily and/or deliberately—such an approach is, of course, suicidal. But if and when we do make mistakes, we ought to make them creatively. Alas, such a prescription is usually far easier said than done for a number of reasons. Essentially, these reasons can be broken down into a simple message: Mistakes are threatening.

Mistakes threaten our position vis a vis our colleagues and business associates and mistakes equally threaten our ability to move up the corporate hierarchy. Senior managers are every bit as captive to the fear of failure and admitting error as the rest of us: They have more to lose, after all. Nevertheless, it does seem strange that organizations normally spend thousands of hours planning an investment and millions of dollars implementing it. And, God forbid, if it does end in a disaster—as many projects do—we do not actively seek to do a "post mortem" for the purposes of drawing appropriate and valuable lessons. Project disasters, more often than not, are simply brushed "under the carpet," as quietly as possible with far too little time spent in critically evaluating and learning from the experiences in order to answer vital questions:

- Was the investment worthwhile?
- Did it go according to plan?
- If yes, how and if not, why not?

THERE ARE LESSONS TO BE LEARNED, IF ONLY . . .

Failures hold morals which successes cannot, especially if we want to go beyond the existing state of the art in project management. Creative engineers constantly seek to employ new concepts to reduce the weight and/or quantity of valuable materials in order to shrink the total cost; in effect, to strive to do more with less through more efficient use of materials. However, each new novel and untried concept or experiment is in the form of a trial leading either to failure or to success. Such is the nature of science and engineering, as with all human endeavors (Petroski, 1985).

Among the earliest engineering structures, for example, is the Egyptian pyramid which was probably designed by a process of trial and error. Its shape is an extremely stable one. Excavations, however, show some of the earliest pyramids having design and shape changes, literally in the middle of construction, as though the builders perceived the instability in the initial design. As another medieval example of learning from failures, consider that when the Cathedral at Beauvais, France suffered a major collapse, it proved to be a turning point in the development of Gothic structures. The designs thereafter became more conservative.

The failure of some bridge structures (e.g., Tay Bridge (1879), Wuebec Bridge (1907), and the Tacoma Narrows (1940)) undoubtedly was a matter of great concern. But let us remember that design choice is much like choices in every walk of life. Several different designs and choices may fulfill the requirements, but in the ultimate analysis, only one can be chosen, just as only one route can be taken on a single trip. And in spite of the best available information and calculating seemingly every alteration, this choice may turn out to be wrong!

Clearly, recklessly courting failure in the hopes of learning valuable lessons in design engineering or product marketability is not the answer either. Failure carries a societal price, sometimes hidden but often visible and profound. For example, one of the greatest structural tragedies in U.S. history occurred in 1981 at the Kansas City Hyatt Regency Hotel in which an overhead walkway collapsed with resulting casualties of over 100 killed and 200 injured. The circumstances, which affected in one way or another practically half of Kansas City's population, are generally well-known. They involved the catastrophic collapse of two suspended walkways crowded with viewers onto the even more crowded dance floor below. Twenty months of investigation found no evidence of a federal or a state crime, but two months later, the Attorney General for Missouri filed a complaint against the engineer responsible for the design of the hotel, charging him with:

- gross negligence in the design/analysis of the walkways, and
- failure to calculate the load capacity of bridge rods and connections.

The systematic study of past project failure is designed to yield valuable data in relation to future projects. Because the cost of such mistakes is usually painfully high in terms of dollar losses and, even worse, loss of life, case studies of project disasters can be an invaluable source of information and provide a real insight into the way in which human error can totally negate our plans and projects.

Words of Wisdom

As a prelude to our discussion of failure, consider what some notable individuals have had to say on the subject:

> *"Worst-case forecasts are always too optimistic."*
>
> (Augustine)

> *"I have learned throughout my life as a composer chiefly through my mistakes and pursuits of false assumptions, not by my exposure to founts of wisdom and knowledge."*
>
> (Igor Stravinsky)

> *"There is the greatest practical benefit in making a few mistakes early in life"*
>
> (T. H. Huxley in *On Medical Education*)

> *"Failure is the normal, natural way of mapping the unknown.*
> *Success eludes you because . . . you haven't failed enough.*
> *People who willingly risk failure . . . have the best chance of success.*
> *Everyone gets a chance. No one lives a failure-proof life forever.*
> *A stiff dose of misfortune is often a painfully effective tutor.*
>
> (Vic Sussman)

> *"Continuous success builds arrogance and complacency . . . I want people who love the battlefield, people willing to risk failure—You learn a great deal more from what doesn't work than from what does."*
>
> (H. Ross Perot)

> *"Everyone in business makes mistakes. If you didn't take risks, you'd never get anything done. The sin would be if you made the same mistake more than once."*
>
> Michael Eisner, Chairman and CEO, Disney Corporation, (quoted in Henkoff, 1995)

> *"A project has but one chance of success."*
>
> (Gordon Gale)

TO ENGINEER IS HUMAN

This section is named for the intriguing title of a book dealing entirely with failure despite so-called "successful" designs (Petroski, 1985). The main message of this book suggests that the impulse to build and to rearrange inanimate nature is as much a part of us as is our instinct to walk and to dance. But just as we occasionally falter in our step, engineers sometimes err in their design.

Included as some of the infamous examples of catastrophic structural failures which have shaken our confidence are the following. Each prompts a basic question: Do engineers always understand their own creations? In 1940, the Tacoma Narrows Bridge, near Seattle, twisted itself apart. In 1979, a DC10 Jumbo jet crashed in Chicago, and in 1981, an elevated Kansas City Hyatt Regency Hotel walkway collapsed. Petroski's book helps explain how dramatic engineering failures can occur even in an age of computer-aided design. It seems that successful feats of engineering have always served as challenges to build longer, higher, lighter, and still more ambitious structures—until a tragic failure occurs due to overlooking some detail or exceeding an unrealized limit. Unfortunately, such failures are part of our inherited culture and our daily lives. Further, these failures, if carefully analyzed and studied, often do more to advance engineering knowledge than all the successful machines and structures. Indeed, failures appear to be inevitable in the wake of prolonged success, which encourages a sense of invulnerability, resulting in concomitantly lower margins of safety. Failures, on the other hand, lead to greater safety margins and hence, new periods of success, with the cycle being inexorably repeated.

The object of engineering design is to obviate failure, but the truly fail-proof design is chimerical. Engineers do their best, but in the ultimate analysis, all designs are inherently "risky" and departures from traditional design are more likely than not to hold surprises. Good design minimizes the effects of surprises by anticipating troublesome details and by overdesigning for an extra measure of safety. Sir Alfred Pugsley, a pioneer in the study of metal fatigue in military aircraft wrote:

> *"A profession that never has accidents is unlikely to be serving its country efficiently."*
>
> (quoted in Petroski, 1985)

Innovations involve calculated risk, some of which are successful, making their engineers heroes, while others are failures and their engineers often condemned. The first edition of *The Life of Reason,* makes this point eloquently in quoting Santayana (p. 284): "Progress, far from consisting in change, depends on retentiveness . . . Those who cannot remember the past are condemned to repeat it."

WHY DO PROJECTS FAIL?

Failure in real life is more frequent than we would like. The trick, whether in managing projects, personal relationships, or professional commitments, demands that we explore ways to cope with it and how to learn from it. This text will suggest some guidelines dictated by (frequently) bitter experience, such as:

- Learn failure (from others' mistakes) instead of "practicing" it yourself.

- Learn to understand it and thereby, learn to avoid it.
- Know your project; assess its weaknesses, both generic and specific.
- Continually reassess your objectives, adjusting them whenever necessary.
- Accept some failures, realizing that this is the price of success.
- Learn from your mistakes; compile lists, update them, and analyze them.

Efforts directed toward innovation, including all new product introductions, typically come out of a project management process of one form or another. Hence, many organizations are often literally "betting the company" on their ability to successfully plan for, execute, and introduce projects into the marketplace. Boeing, for example, will routinely commit hundreds of millions of dollars in research and development and related expenses to a new airframe design before receiving their first orders. Against such a backdrop, it is clear that the stakes in successful project execution are extremely high. To illustrate, consider the case of the development, manufacturing, and marketing of a pharmaceutical product. Consider the implications, demonstrated by a simplified flow diagram of the development process:

Process	*Benchtop*	*Clinical*	*Pilot*	*Commercial*	*FDA rev/Marketing*
Phase	I	II	III	IV	V
Time in months	18	6-12	6-18	12-60	12-36
Cost, in dollars	0.5-5.0	———	5-100	———	3-100
Success, % Prob.	5-20	70	30	25-30	1-5

The stakes are high indeed and the probability of success is rather low, both at the start and towards the end. Planning must always take into account the ultimate goal in terms of what is needed to obtain marketing approval. Once the final goal is set, one works backwards to complete the entire development plan. However, in many companies, there are time and cost overruns at various stages due to breakdowns in communications and lack of cohesion in the entire process. The cost and time requirements increase progressively and somewhat exponentially with each successive phase. In view of such implications it is absolutely vital to carry out additional studies at an early phase to avoid discovering that a crucial premise was incorrect or faulty.

To improve the chances of success, product development must be treated as a dynamic process. Too often, companies latch on to an initial strategy that may have become negated due to any number of external events. Consequently, managers must realize that sticking to the so-called "terrific plan" to the bitter end is likely to lead to disaster. The situation must be reviewed periodically and appropriate changes in the master plan affected. Good science implies that every stage of the company's development plan must be carried out with rigorous scientific and ethical integrity. From time to time it is imperative to ask even naive questions such as: Would you invest your money in this project? A negative answer to such a question is a strong indicator of the need to reevaluate the position and decide on corrective action, perhaps even a detour.

But such mid-course corrections are often harder than they seem. In an intriguing recent book, Robert Tietelman (1994) demonstrated that learning from history is much harder than writing about it. The author provides an excellent summary of success and failure in business in the U.S. since the 1930s. One point he demonstrates time and again is the inability of businesses to take to heart lessons and mistakes from the past, in particular, *their own* past. Case after case is cited of organizations that have failed through repeating correctable errors, often several times.

EVEN THE MIGHTY STUMBLE OR WANTED: A NEW BREED OF MBAS

Legendary corporate giants such as GM (automobiles), IBM (computers), RCA (consumer electronics), and XEROX (office automation) have failed to compete with their competitive counterparts in recent years, both in domestic and foreign markets. Why?

Clearly, the answer to such a question is multifaceted, calling upon attention to the conceit of past success, the complacency of large size, and the unwillingness to swim against the tide of corporate bureaucracies. We suspect however, that in most cases it was not because they lacked the technical means to succeed, but rather the ability to manage their technical strengths so as to convert them into an abiding competitive advantage. In effect, the travails of these companies are often management, not technology failures. Does management education need a major shift?

It would appear so. In support of this, we mention a somewhat hidden fact that in Japan (as well as in France and West Germany), the number of MBA graduates per year is less than 100, in sharp contrast to some 80,000 in the U.S. A somewhat similar picture also seems to exist in respect to the number of qualified accountants. This disparity is so great that it warrants a serious study including its effect on success in business in general and industry in particular. The subject is somewhat outside the purview of our present effort and we must leave it at that. It does seem that in its current form, the traditional MBA education may have outlived its utility and there is need for some serious soul-searching in this regard.

The present traditional MBA curricula has been severely criticized in recent years. We refer here to just two recent studies in this regard. The first (Gupta, Naik, Paranjpe, and Kumar, 1995) is based on a survey of senior managers, asking them their opinion of modern MBAs. The study sample comprised 1,382 respondents, of whom one-third were MBAs and the rest non-MBAs. Nearly 75 percent of the respondents (both the MBA and non-MBA segments) believed there is no difference between MBAs and non-

MBAs. Among the perceived shortcomings of MBAs are knowledge-base and interpersonal relations.

The other article (Hawkins, 1994) describes a recently introduced management and technology (M&T) program at Rensselaer Polytechnic Institute (RPI) designed to invest the core management curriculum with a strong technological orientation. This new course of study leading to an MBA degree covers not only management and technical disciplines, but also the interface thereof and thus is intended to bridge the major gap:

Management schools: Graduates who do not understand and cannot influence technologists.
Engineering/Technical schools: Graduates who do not understand and cannot influence management.

In fairness to American business schools, they are recognizing at an increasing rate the need to revamp their curriculums to reflect the new paradigmatic shifts in business education. The focus is becoming increasingly international, while the old and traditional majors are giving way to a more rounded program of study, aiming at creating stronger interpersonal and cross-disciplinary skills. These changes are expected to lead to a new breed of managers, a cross between the sorts of managers and engineers held in such urgent need today. Unconsciously perhaps, Japan already has adopted this cross-fertilization with an added dimension of finance. The latter is done by encouraging engineers to become accountants as well or at least learn the basics of accounting, thus solving an otherwise serious problem—the lack of communication (for want of a common language) between engineers and accountants.

Clearly, the teaching of business education is in a state of transition, as is our general business environment. Competition at global levels threatens almost all industries, particularly when they have allowed themselves to grow complacent along the IBM model. This book is intended to serve as the sort of "jolt" that some of us need in our linear, programmed thinking about the conduct of business. Rather than recite a string of successful project examples (certainly useful in their own way), we have sought a less-traveled road: Attempting to drive home the lessons of pragmatic and effective project management through an honest appraisal of what has not worked. Our own past work and consulting experiences enable us to feel confident that each reader will find something of themselves and their own organizations in these pages. That sort of recognition is our goal. If the lessons so painfully gleaned by other organizations can save a future firm from following such a path, this book will have been well worth the effort put into it.

References

Brooks, F. 1995. *The Mythical Man-Month: Essays on Software Engineering,* 3rd Ed. Reading, MA: Addison, Wesley.

Gamerman, G.E. and B.F. Mackler. 1994. Winning in today's biopharmaceutical marketplace," *Chemtech,* vol. 24 Dec, pp. 37–41.

Gordon, G. 1991. "Managing capital projects successfully," *Management Accounting,* 63, March, pp. 59–64.

Gupta, T., Naik, S.C., Paranjpe, A., and Kumar, G. 1995. MBA's vs. non-MBA's, *Times of India,* Bombay, Jan 19, ASCENT supplement, p. I.

Hawkins, J.P. 1994. "A new breed of MBA," *Chemtech,* vol. 24, Dec, pp. 10-13.

Petroski, H. 1985. *To Engineer is Human—The Role of Failure in Successful Design.* New York: St Martins Press.

Sussman, V. 1994. "Don't Fear Failure," *Readers Digest,* Jan, pp. 47–49.

Henkoff, R. 1995. "Smartest and dumbest managerial moves of 1994," *Fortune,* Jan 16, p. 59

The Life of Reason, Vol 1 page 284

Tietelman, R. 1994. *Profits of Science—The American Marriage of Business and Technology.* New York: Basic Books.

Willigan, G.E. 1989. "The case of the expensive expansion," *Harvard Business Review,* vol. 67 Jan-Feb, p. 10+, 8p.

2

*Projects Are Not Products**

PROJECTS ARE *NOT* PRODUCTS. The two are different in some very important ways. These differences need to be examined prior to our discussion of project failure so that readers can clearly distinguish the unique characteristics that define project-based work. While there are a number of definitions of projects, we prefer two that have been advanced by Steiner and Cleland and Kerzner. The first, by Steiner (1969, p. 498), suggests:

> *"A project is an organization of people dedicated to a specific purpose or objective. Projects generally involve large, expensive, unique, or high risk undertakings which have to be completed by a certain date, for a certain amount of money, within some expected level of performance. At a minimum, all projects need to have well defined objectives and sufficient resources to carry out all the required tasks."*

---

* Portions of this chapter were adapted from Pinto, J.K. and O.P. Kharbanda, *Successful Project Managers: Leading Your Team to Success*, New York: Van Nostrand Reinhold, 1995.

The second definition is offered by Cleland and Kerzner (1985, p. 199) in their work, *A Project Management Dictionary of Terms,* and includes the following characteristics:

> *[A project is] "A combination of human and nonhuman resources pulled together in a temporary organization to achieve a specified purpose."*

Using these and other definitions, it is possible to isolate some of the important characteristics underlying projects. Most writers on project management point to four common dimensions of projects:

- They are constrained by a finite budget and time frame to completion; that is, they typically have a specific budget allocated to them as well as a defined start and completion date
- They comprise a set of complex and interrelated activities that require effective coordination
- They are directed toward the attainment of a clearly-defined goal or set of goals
- To some degree, each project is unique

These features form the core that distinguishes project-based work from other forms of organizational activity. Because they are significant and underscore the inherent challenge in managing projects, let us examine each of these characteristics "up close."

FINITE BUDGET AND SCHEDULE CONSTRAINTS

Unlike the typical ongoing operations that occur within functional units, projects are set up with two very important bounds on their activities: a specified time period for completion and a limited budget. Projects are temporary undertakings. They are not intended to supplant the regular functional operations of the organization but rather, operate until they have achieved the goals they set out to accomplish. Once these goals have been completed, the project ends. Of course, budget and time constraints are generally estimates, based on the best available (and sometimes naively optimistic) information that the organization has. As a result, it is not uncommon to build in a margin for error to allow for unforeseen expenses or time slippages. The key to understanding the nature of project work as opposed to, say, an assembly line production run, is that unlike the production line, which can continue on into the indefinite future, projects are temporary. They fulfill their goals in accordance with time and money limits, and disband.

Complex and Interrelated Activities

Projects typically comprise a degree of complexity that is not found to the same degree within other functional departments, often due to the cross-functional nature of the activities. For example, in developing a new product, a project team may be staffed with members from a wide variety of functional backgrounds: marketing, production, finance, human resources, and so forth. This cross-disciplinary nature of much of project-based work adds another order of magnitude to the usual levels of complexity found within any individual department.

Unfortunately, the complexity of projects and interrelatedness of team members leads to an unwelcome side effect: conflict. Sources of conflict abound within projects due to the unique properties that they possess as well as the multiple goals and attitudes of different members of the project team. Not only are projects forced to compete with functional units for their share of a variety of scarce resources, but even within the project team, the almost ubiquitous nature of conflict is clearly demonstrated. Members of the project team regularly experience these sources of tension and must seek a balance between dual allegiances to their functional bosses and project manager.

Another example of the complex and interrelated nature of projects derives from the multiple activities that are carried out, often simultaneously by different members of the project team. This interrelatedness is typified by the scheduling techniques of PERT and Gantt charting, which demonstrate the sometimes bewildering array of interdependent activities performed by a project team. If tasks are not performed in their correct order and within the time allotted to them, the entire project can be jeopardized. Consequently, interrelatedness implies two distinct challenges: First, the need to establish a productive working relationship among team members from diverse functional backgrounds and second, to create a scheduling and control system that permits the greatest level of efficiency of project activities.

Clearly Defined Goals

Projects are usually created with specific purposes or a narrowly-determined set of purposes in mind. Indeed, the worst sorts of projects are those which are established with vaguely defined or fuzzy mandates that permit a wide range of interpretations among members of the project team and parent organization. Projects of this sort are usually doomed to spin along out of control as goals are continually interpreted and reassessed while the budget grows and the estimated completion date slips further and further into the future. The more well-defined the goals, the clearer the indications are, both internally and externally, that the project team is succeeding. We often find that one of the chief features of projects that continue to "function" well past the point of serving any reasonable purpose is that they represent those for which the initial goals have either been altered mid-stream or were so poorly elucidated when the project began that they provided no guidance for the team.

Uniqueness

Projects are usually "one-shot" deals; that is, they are non-recurring and typically established to address a particular problem or market opportunity. Their very uniqueness is the characteristic that underscores the challenge of project management—the learning curve from one project to the next is, at best, tenuous. Because we are faced with a unique problem or opportunity, the "rules" for how the project should be configured and run have not been developed. In effect, we have to learn some lessons as we progress. In learning these lessons and exploring virgin territory, project managers encounter the sort of risks and uncertainty that typify project-based work. It is, however, important to note that a project's "uniqueness" may vary to a considerable degree from company to company and project to project. Consider a construction company specializing in building high-rise apartment complexes. While the designs are different, the basic process is such that it may be possible to take institutional learning from one site to the next. On the other hand, a company such as Fluor-Daniel, which specializes in unique industrial construction projects, is often faced with having to relearn various and challenging complexities that become part of each job.

While projects are unique, one-time processes, product manufacturing tends to repeat ad infinitum. And yet, in a sense, there is a connecting thread. Projects precede the product; that is, the process of product development via bench scale research and development and pilot plant operation is typically thought to be a project undertaking. And of course, the cycle is completed by having new projects which often lead to innovative products—as we demonstrate here.

Time and cost overruns are, unfortunately, the norm rather than the exceptions that they ought to be. There are other factors, such as politics, changing technology, and the economic climate, which are largely outside the control of the project management team, which also play a significant role. However, in essence, the control of cost and time is a primary project management responsibility.

Project Management vs. Line Management

Projects and project management are seemingly as old as history itself. In the ancient world, we are treated to examples of the sorts of successes mankind is capable of in managing its projects correctly (e.g., Egyptian pyramids, Stonehenge, the Taj Mahal, China's Great Wall). In spite of centuries of experience behind us and Bronowski's philosophy of each generation standing on the "shoulders" of previous generations, managing projects has gotten no easier, despite the fact that there have been perhaps several hundred books published on this and related subjects in recent years. Why then, is project management so hard? Simply, because it is one of the most complex undertakings in which organizations can engage. The interrelatedness of its operations combined with the unique nature of each project makes it extremely difficult to offer project managers more than a rudimentary roadmap for finding their way through the minefields of disaster in order to achieve project success.

Project management is difficult precisely because it is characterized differently than simple functional organizational activities. It upsets the status quo, it threatens comfortable methods of operation and product line manufacturing, it encourages creativity and operation "outside" the bounds of the traditional hierarchy, all while requiring the project leader and team to keep a weather eye open for schedule and budget constraints. Project management is enormously difficult because by definition it represents change and challenge, two ideas that most organizational members find threatening. If done well, its successes are spectacular. If done poorly, its failures are monumental.

What of R&D projects? R&D relates to areas unknown, unless of course (and this is happening far too often, unfortunately) we are seeking to "rediscover the wheel." Unknown means exactly what it says. We cannot delineate a clear route to our final goal, often because beyond sketchy phrases, we may not know precisely what the final goal comprises. Hence, to use an example from the chemical industry, our goal may be as simple as researching and developing a process for making carbaryl other than via the MIC (the notorious Bhopal culprit) route. We know the chemical composition of the product and possible routes to it, but that is about all. Part of the project development process is about basic research, followed by development, then the creation of a pilot plant. However, each of these areas is a project by itself. Starting with maximum uncertainties, we will hopefully be able to reduce the various choices/options as we go along.

Project management differs from the conventional line management in many ways but particularly, in one most vital respect: A project has but one chance of success, whereas with conventional line management there is always the opportunity to "do things better next time." (Kharabanda and Stallworthy, 1984) Failure to recognize this vital difference has been one of the main factors in project failure. Management philosophies and styles are necessarily different in the two cases. In the case of line management, because we are faced with recurring tasks, it is possible to grow used to the processes, learn their strengths and weaknesses, and improve incrementally. With projects, the process is unique, the time frame is immediate, and the consequences of waiting and hoping to make improvements at the margins are likely to be highly dangerous to the chances of successful completion.

We propose to demonstrate the mistakes made in project management by taking specific projects and analyzing them. When we study what may be termed a "project disaster," the mistakes made usually seem very apparent, blessed as we are with the benefit of hindsight. This enables us to learn from the mistakes that others have made. It is, however, interesting to note how few companies are willing to engage in their own honest, post-failure project appraisal. The tendency is to avoid discussion of one's failures and it is true that "failed," "bankrupt," "broke," and "bust" are not pleasant words. Yet they are a fact in the corporate world and the number of company failures worldwide is a continuing testimony to their reality. For these reasons we think it will be helpful to spend some time examining a few notable project disasters. This may well help others to learn how such tragedies may be prevented, and that is the first step on the road to success.

PRODUCT INNOVATIONS ARE EXCITING AND CHALLENGING

Projects are, of course, the "engines of growth" for a company and often, the larger economy. Each project leading to a product, either the upgrade of a conventional, well-known one or a new and innovative product developed through intensive research and development, offers its parent organization the potential for huge profits and a dominant market position, sometimes briefly and other times for years into the future. Project management is a challenging function, but the management of projects for innovative products is even more so. One of the most dramatic differences lies in the nature of cost and schedule estimating. Within the realm of new product development, particularly when the technology is new or the challenges are unique, the process of estimating various time frames and budgets is speculative at best. We will see over the course of this book that many problematic projects or outright failures were often handcuffed by utterly unrealistic scheduling and budgetary constraints at their conception. The inevitable overruns cast negative clouds over the project early on, and in some cases, these poor reputations have lingered throughout the project's existence.

One of the classics in this field, *Managing New Product Innovations,* was written by an authority in this area, Dr. William Souder. This work has been followed by a host of books on the same subject in recent years, (e.g., Martino, 1995; Leavitt, 1994; Hayes, 1994; Burns and Stalker, 1994; Dolan, 1993; Souder, 1993). Souder demonstrates the unique challenges of managing R&D projects. His central theme shows that R&D is fraught with many uncertainties. In the case of most projects, the team seeks to freeze the technology and performance criteria in place as quickly as possible. Unfortunately, the nature of R&D projects makes such "spec freeze" much more problematic, if not impossible. R&D is undertaken to develop a process, know-how, or technology, either for an existing or a new product, and is by definition, predicated on uncertainty and constrained information.

Souder's 1987 work was the result of ten years of field research on 289 new product development innovations at fifty-three companies. Funded by several agencies, government and private, it is a comprehensive book, also covering ways in which to generate R&D ideas. He characterizes many of these R&D questions as falling into one of three main categories of business problems, including:

- Ideas to meet some stated product or service requirements, sometimes general and other times more specific
- Ideas to solve a particular customer's problem
- Ideas for a product to meet a list of performance requirements

There are additional characteristics of R&D projects that lend them their distinct attributes. Obviously, the most telling feature is that these projects are run within R&D departments or technical functions. Technologists have, by definition, much longer time frames where product development is concerned (Goodman and Lawless, 1993). Left to their own devices, there is a natural tendency for technologists to work to fully solve a problem, to "get it right," regardless of the time spent. Clearly, the constraints of project schedules fly in direct opposition to their natural predisposition. Other distinguishing features of R&D project activities include the finding that, as one wag put it, "Our 'R' is fine, it is our 'D' that doesn't always work." His point was that, within many organizations, procedures for utilizing the collected ideas of the R&D department are seldom well prescribed. Specifically, there is no well-developed process for coding, storing, handling, and disposing the collected ideas. Consequently, a sad but ironic result is that a number of potentially exciting new project ideas are allowed to languish in the organization's "black box," somewhere between research and introduction. Finally, in many instances, specific screening procedures are found to be lacking. Large-scale research has shown that screening meetings are often unstructured discussions among technologists, with a consensus sometimes (but not often) emerging (Souder, 1993). Further, the more technical the discussion, the greater the likelihood of lapsing into the state that Aesop warned of long ago:

Beware that you do not lose the substance by grasping at the shadow.

(Aesop, 550 B.C.)

Projects fail or succeed due to a variety of complex reasons. For either failure or success, there is a "constellation" of factors, but the complexity arises from the fact that the absence of even one important factor from the set can make all the difference in the outcome (Souder, 1987). Even seemingly extraneous factors like new laws, changes in consumer tastes, shifts in demands, and changes in needs can make an enormous difference in the fate of a project—overnight!

Souder's research provides some strange and even somewhat surprising but very pertinent findings:

- Allocating too much money to a project can actually be more detrimental than providing too little money!
- Defensive efforts aimed at protecting a firm's current position do not necessarily lead to high success rate. This has important implications for companies that seek to reduce R&D risk by sticking with tried and true techniques or technologies.
- Beware of the common myth that once the "hard" research work is done, developmental project management will be easy; this can very easily lull the managers into intellectual complacency.
- Despite all the publicity suggesting the contrary, customers are not always a company's most fertile source of new product ideas. In-house ideas, particularly from the marketing personnel, are particularly successful.

As expected, a close developer-customer interaction dramatically improves the chances of product/project success. This includes contacts at personal levels, as well as at the industry association levels. The benefits from these also extend to evolving the new product standards, buffering domestic competitors, and even forestalling the entry of foreign products and competitors.

Some additional points that should be clarified include the myth of "deep pockets." The argument is frequently raised that larger corporations have an advantage in new product development due to their size and access to resources (principally money). For example, it has long been a point of pride with IBM that they devote 10 percent of sales revenue to R&D. In fact, in 1991, that amount totaled $6.644 *billion* dollars, giving IBM the largest and certainly costliest private R&D labs in the world. Yet, at the same time that IBM was spending huge amounts of money on basic research, they were unable, due to bloated bureaucracies and other reasons, to get products to market quickly. In fact, they frequently lagged the field in their industry. Actually, such stories are not surprising: Tom Peters (1987) reported a National Science Board finding in which only 34 percent of major technical innovations come from giant firms (over 10,000 employees), far less than their share of industrial output. Further, the smallest firms produced about four times as many innovations per R&D dollar as the middle-size firms and twenty-four times as many as the larger firms!

Also contrary to the popular conception, large companies seem to carry out a disproportionately small share of the R&D aimed at entirely new products and processes. Further, small firms spend three to ten times more than large ones to develop similar new products. Even in the case of DuPont, the bulk of the firm's commercially important products have historically been invented *outside* the firm.

THERE IS MANY A SLIP 'TWIXT THE CUP AND THE LIP

Despite the most innovative products and the best project management, success may elude even one of the most singularly successful companies. Let us return to the story of IBM, a name which for several decades now has meant just one thing: computers. This is the case with a company which dominated the computer field for so long as to earn the sobriquet (in relation to its competitors) of Snow White and the Seven Dwarfs (Sobel, 1984; Harris, 1985a; 1985b).

The company, and specifically one of its major projects, System 360, was a singularly remarkable example of a highly successful project. Well, times have changed! In recent years, due to a combination of factors, the Snow White image has been steadily shrinking while the dwarfs have been getting bigger and bigger, rendering the fairy tale story title no longer valid. How did this occur? Consider their successful development of the personal computer and the subsequent missteps they took.

IBM developed its first PC in thirteen months flat, a remarkable achievement. Under time pressure to beat the competition and to gain an enormous headstart, the thirteen-member PC team was specially empowered to take decisions that in normal course would neither have been considered nor approved. For example, the PC team went outside IBM to procure all the components on strict basis of price, quality, and delivery. IBM companies were, of course, like any other supplier, allowed to bid and expected to prove their merit in addition to meeting fully the team's requirements. In the course of project execution, the team discovered what was later termed "synchronicity." This meant that to gain time and a head start in this fast growth sector, software was being developed well before the hardware was perfected. To top it all off, the project chief, Don Estridge, knew perfectly how to motivate his team members and how to keep their morale high.

Well, it was all worth it, and the plan worked to near perfection. And yet, the company failed to latch on to the enormous headstart it gained over its competitors. What happened? It is far too long and complex a story to bring you all the details, and we must once again rely only on the headlines. According to Heller (1994, jacket flap):

> *"Who killed IBM? . . . Management? Competitors? Market? . . . The multibillion dollar losses of the early 1990s started with the ousting of*

> *John Akers as chairman . . . [T]he richest and the most powerful corporation . . . threw away . . . the advantages it had built . . . over many decades . . . [T]he administration, the high-church mentality of the management, and the salesmen-driven dominance . . . [coupled with the] personal computer . . . achieving runaway success . . . IBM's inertia and lack of innovation . . . was exposed when [its] major market—in mainframes—began to dwindle . . . The human and technological disaster . . . [leading to] a sensational fall . . . [—] from virtual monopoly to financial woe.*

Was IBM's fall from grace understandable in retrospect? Perhaps. But it illustrates precisely the message of our present effort—failure and disaster, especially in the project field, hold a number of lessons for management provided we are willing to shed the facade of hubris and understand that IBM's crash could merely presage those of other, seemingly invulnerable firms.

As one author, the chairman of a large marketing consulting firm, has noted in writing about IBM (McKenna, 1994):

> *"IBM is an obsession . . . [a] monolithic giant . . . [T]he competitive forces that made IBM the powerhouse of the 1960s and 70s no longer work in the 1980s and will not in the 1990s . . . Using technology as a driving force . . . Apple, Compaq, Cray, Digital . . . ignored the power of Big Blue . . . [T]heir innovations broadened, expanded . . . creating whole new . . . users. As a result the slow-moving giant is now reacting to change rather than setting its agenda."*

INTEGRATING R&D AND MARKETING

It is clear that marketing is the real driving force of R&D and indeed of the entire manufacturing center. For best results there must be close integration of the various sectors. How this is managed in practice by American and Japanese companies has been the focus of many studies. One such intensive study (Norton, Parry, and Song, 1994) has analyzed survey data collected from R&D and marketing managers working for American and Japanese chemical manufacturers. Three key differences have been found:

- In Japan, marketing seems to have a somewhat narrow role (that of information resource) in respect to its role in new product development.
- In Japan, R&D and marketing have an asymmetric relationship in new product development, with R&D playing a dominant role.
- In America, there exists a greater potential for conflict between these two functions.

There was clear evidence that R&D departments in Japan play a larger role in new-product development than do marketing departments. American firms, on the other hand are not characterized by such asymmetry in their relationship. In addition, the American firms included in this study showed a greater potential for "turf battles" between R&D and marketing functions.

Thanks to the prevalent management practices in Japan, managers are reassigned to various functional areas, as a matter of routine, during the course of their careers. This helps to inculcate in the managers skills and perspectives of other functions. This is a major factor in encouraging cross-functional cooperation. In many American firms, such job rotation is seen as an additional cost factor, but its absence clearly contributes to perpetual conflicts.

Another difference between the two is that Japanese managers place greater emphasis on legal and regulatory information. It may be that they are particularly sensitive to legal issues because they operate in a dominant export economy. This is reflected in sharply different attitudes. Ask them about their market shares: An American manager will usually think only in reference to the home market, whereas his Japanese counterpart thinks only in terms of the global market share. In general, Japanese managers are far more information-hungry than their American counterparts. Thanks to their much wider horizons, Japan has maintained a competitive edge in balance of trade economics, products, and projects.

MANAGING DEVELOPMENT PROJECTS

To win in the global competition, companies need to develop more successful products than ever before. The important issue here is not necessarily more advanced products as much as a steady stream of continuously improving products. Shortened life cycles and dramatically narrow market "windows" frequently make it costly and ineffectual for companies to seek to hit the "home runs" in relation to their new product offerings. Rather, the key lies in steady innovation, coupled with a long-term outlook. And the more innovative the product, the more competitive companies become. Vincent (1988) has underscored the challenges of the innovation process, arguing persuasively that innovation is the lifeblood of the manufacturing industry, and lately, of service industries as well. Success is now increasingly dependent on understanding and managing new technologies and using it extensively to improve competitive edge. As Vincent notes: Apart from setting clear goals and the right management approach, it is absolutely necessary to review progress periodically. Above all, technology must always be seen merely as a means to an end and not an end in itself.

In order to do things in a hurry, companies often undertake far too many development projects at a time—a common pitfall. The limited resources of men and materials get spread out so thinly that not much gets achieved. Such a situation is more likely to happen if the company does not have a formal process for choosing and managing its development projects. What is required is a process for mapping out an "aggregate project plan," with concentration and attention on a few development projects of strategic importance (Wheelwright and Clark, 1992). To achieve this requires drastic action. The development process must be reformed by eliminating the lion's share of existing projects and picking only those which are vital to the company's core business. Such action in the case of one company led to a reduction in the number of development projects from thirty to eleven, thus eliminating nearly two thirds of the extraneous projects. Final selected projects were well defined and strategically positioned within five broad platforms:

Project Type	Number of Projects
R & D projects	3
Breakthrough projects	1
Platform projects	3
Derivative projects	3
Alliance/partnerships	1

It was through use of a similar approach that Sony pioneered the "hyper-variety" strategy with its over 200 Walkman models based on only three broad platforms, thus making the entire job easily manageable. In conclusion, Vincent (1988) offers eight steps in an aggregate project plan:

- Define project plans under the above five categories
- Identify existing projects and classify them under these categories
- Estimate time and resources required for each based on past experience
- Identify existing resources available
- Determine the desired mix of projects
- Identify projects which cannot be accommodated within existing resources
- Decide which specific projects to pursue
- Work to improve development capabilities and concentrate on final projects

Arriving at an aggregate project plan is not easy, but the effort is well worth it. And the results will speak for themselves. Of course, there should be a built-in flexibility to permit changes as events unfold and managers make the necessary adjustments. But the above process will provide a solid foundation which can serve as a sound basis for a company's future and its growth.

References

Bronowski, J. 1973. *The Ascent of Man,* London: BBC.

Burns, T. and Stalker, G.M. 1994. *Management of Innovation.* Oxford, UK: Oxford University Press.

Dolan, R.J. 1993. *Managing the New Product Development Process: Cases and Notes.* New York: Addison-Wesley.

Harris, M. 1985a. "IBM—More worlds to conquer," *Business Week,* February 18, pp. 84–7.

Harris, M. 1985b. "IBM graphs fanatical drive to beat Japan in manufacturing," *Business Week,* March 11, p. 62.

Hayes, W. 1994. *Total Quality in R & D.* New York: St Lucie Press.

Heller, R. 1994. *The Fate of IBM, 1994?* New York: Little Brown.

Kharbanda, O.P. and E.A. Stallworthy. 1984. *Project Disasters with a Moral for Management.* Aldershot, UK: Gower.

Leavitt, J.S. 1994. *Total Quality through Project Management.* New York: McGraw.

Martino, J.P. 1995. *R & D Project Selection.* New York: Wiley.

McKenna, R. 1994. *Who's Afraid of Big Blue?* New York: Addison-Wesley.

Norton, J., Parry, M.E., and Song, X.M. 1994. "Integrating R&D and Marketing—A comparison of practices in the Japanese and American Chemical Industries," *IEEE Transactions on Engineering Management*, 41, pp. 3–20.

Peters, T.J. 1991. *Thriving on Chaos—Handbook for a Management Revolution.* New Delhi: Tata McGraw-Hill.

Sobel, R. 1984. *IBM—Colossus in Transition.* New York: Sidgwick and Johnson.

Souder, W.E. 1987. *Managing New Product Innovations.* Lexington, MA: Lexington Books.

Souder, W.E. 1993. *Managing New Technology Development.* New York: McGraw.

Vincent, G. 1988. *Taming Technology: How to Manage a Development Project.* London, UK: Kogan Press.

Wheelwright, S.C. and Clark, K.B. 1992. "Creating project plans to focus product development," *Harvard Business Review*, vol. 70, Mar–Apr, pp. 70–82.

3

What Is Failure?*

INTRODUCTION

Any book that stakes out the analysis of failed projects as its subject, first must contend with the task of stating as precisely as possible, how it intends to evaluate and classify project failure. This process is by no means a simple one, for we will show in this chapter that project "failure," like beauty, is often in the eye of the beholder. Put another way, until we can establish a set of criteria that have some generally-accepted basis for assessing proj-ects, at best we run the risk of mislabeling as failures projects which may in fact, demonstrate the opposite conclusion. This chapter is our attempt to develop a generally-acceptable guideline by which proj-

* Portions of this chapter were adopted from successful information system implementation (1994), by J.K. Pinto. Upper Darby, PA: PMI publications.

ect success and failure must be judged. At the same time, we recognize that it may be difficult for many readers to completely agree with our criteria, as the subject lends itself to a certain degree of passion, particularly among those who have, like ourselves, worked on "failed" projects in the past.

THE UNIQUE SETTING OF PROJECT MANAGEMENT

In the last chapter, we termed projects as the "engines of growth" for most corporations. By this assertion, we suggest that almost all innovative new products developed within companies arise through the creation of projects and the use of project management techniques. Because projects play such an increasingly significant role in organizational profitability, it is vital that we develop an understanding of their unique properties: what is it about the nature of project management that gives it such a preeminent role?

The answer to this question can be made partially clear by demonstrating the position that projects, particularly new product development projects, occupy in our organizations. The global economy, coupled with rapid product obsolescence and shrinking development budgets necessitates a level of competition that is becoming increasingly frenetic. Within many hardware and software computer manufacturers for example, product life cycles have shrunken to mere months—a far cry from a decade ago when the IBM personal computer was the industry standard for nearly five years! Coupled with this increased competition is the concomitant demand for innovation as companies vie for transitory market dominance. The most obvious method for "leading the field" in a new market opportunity lies in rapid product development, hence the need for professional and competent project management practices.

Project management offers a number of advantages to companies but it also creates a series of special challenges. For example, typical new product development lies outside the realm of traditional departmental budgeting and control practices. Upper management is often forced to stake tremendous amounts of capital and other resources on the mere promise of a good idea, in the hope that an innovative product will eventually emerge. Consequently, with many large projects, initial budgets and schedules comprise best guesses as to the most likely costs to be incurred. It is no wonder then that we continually read about projects saddled with runaway budgets and schedules sliding, as initial plans meet hard reality.

Another difficulty within the project management context refers to the lines of responsibility for development. Most projects are staffed with personnel borrowed (sometimes on an extended basis) from their line departments. A project manager faces the unique and vexing challenge of attempting to motivate and secure the commitment of a disparate group of functional employees who often feel little personal identification to the project and preserve a profound functional loyalty. This problem is further exacerbated since in most companies the project managers often do not have the freedom to perform formal performance appraisals on these "temporary" subordinates.

When set against this unique backdrop, project control becomes an increasingly problematic issue. The irony is that most organizations, including those that have experienced terrible difficulties with their projects, will readily acknowledge the necessity of using project management in their operations. What is particularly frustrating to them is that they continue (seemingly) to make the same mistakes again and again as one project after another falls further and further behind, goes increasingly overbudget, and then is either canceled outright or simply slides into oblivion. In our experience, the problems that underscore such shaky track records are usually legion; however, one common denominator shared by most troubled organizations is their unwillingness or inability to learn from past failure. Project review meetings, so important as a learning tool, are either perfunctory (in the case of successes) or ignored (in the case of failures). This is a pity and it is with this message in mind that we have developed this book as a learning guide for failure. Somewhere buried in the stories to be told in these pages, there are specific lessons for almost all readers. It is our hope that each of us is strong enough to recognize the dross of common experience as readily as we are to claim the laurels of success.

DETERMINING SUCCESS AND FAILURE

It is, perhaps, ironic that attempts to label certain projects as "failures" often pose considerable difficulty. Part of the problem, we suspect, is that while everyone seems to understand intuitively the various component measures of project "success" or "failure," agreement with how to assess various projects in light of these criteria is often problematic. Put more simply, ugliness—as with beauty—is a highly subjective assessment. Some of the most impassioned arguments we have received from project managers in our consulting and training careers have come about as a result of our terming a specific project with which they were associated as having "failed." This response is completely understandable in that project managers' careers often hinge on their ability to "deliver the goods" in the form of successfully completed projects. Consequently, it is axiomatic that in the absence of dramatic disaster (e.g., structural collapse in construction or banned or abandoned pharmaceutical development), it seems for every detractor of a specific completed project there is often a champion singing its praises.

How is this possible? How is it that two equally capable individuals can often view the same results through entirely different frames of reference and make completely opposite conclusions about a project? Part of the answer lies in the notion that project success is not always as clear-cut as we would sometimes believe. Any one of a number of confounding issues can cloud our ability to view a project's outcome in any sort of objective light. For example, the point in time when a project is evaluated can make a very real difference in its evaluation. Likewise, egos and personal agendas of top managers in a company can serve to obscure the true outcome of a project, as these powerful individuals seek to protect themselves and their turf from the side-effects of bumpy projects.

Finally, it is often the case that while successful projects are trumpeted throughout the organization and publicized externally, the majority of project failures are quietly swept under the carpet. People naturally tend to promote the positive and wherever possible, adopt a simple philosophy: Out of sight, out of mind. The irony, of course, is that all organizations experience project failure far more often than rousing success. Consider, for example, the results of a recent study by Peat Marwick of 300 large companies attempting to implement computer software or hardware development projects. Sixty-five percent of the organizations reported experiences where their projects were grossly over budget, far behind schedule, and/or the technology was nonperforming, leading to "runaway" projects. In some cases, the companies experienced all of the above! Perhaps more impressively, over half of these firms considered this state as "normal" or "of no concern." (Cringley, 1994)

Because of the confusion surrounding the meaning of project failure, we cannot begin making meaningful observations about the success or failure of a particular project (as the balance of the book intends to do!) until we develop as systematically as possible, a working definition of project success (and by default, failure) that we can use from this point. We are under no illusions that this exercise will reconcile the various views and arguments of all readers—too much ego and potential career payoff are involved on the part of some. Nevertheless, such a framework will at least allow us to develop a foundation upon which to build the arguments we intend to advance while illustrating the various and distinct ways in which projects can fail.

What, then, is an appropriate definition of project success? In the old days, project managers commonly made use of a concept known as the "triple constraint" to evaluate a project at completion. This triple constraint offered a three-legged stool for any project's viability. The three constraints were: a) time—the project had to come in on or under its initially scheduled timeframe, b) money—the project had to be completed within its budget limits, and c) performance—the end result had to perform in the manner that was intended. Seen in this light, it was relatively easy to make some initial value judgements about any project. One had only to consult the project's timeline to assess schedule constancy, review the cost accountant's report to determine budget adherence, and see if the project "worked" to measure performance.

Unfortunately, while simple, the triple constraint has also become increasingly simplistic in the modern business world. In an era of tremendous competition and enhanced concern for customers, the triple constraint has become a dangerously out-of-date convention. In considering the three components of the triple constraint, it is clear that the primary thrust of each of these measures is *internal;* that is, they are intended to satisfy some interest group internal to the organization rather than in the environment. For example, satisfying time and budget considerations are often the concern of cost accountants who are tasked with keeping costs down. Likewise, the performance criterion has often been seen as primarily an engineering concern—the challenge of making a product that works.

What was lost in the confusion was any real concern for the customer; that is, the desire to satisfy the concerns of the client for whom the project was intended. Within many companies with which we have worked or consulted, a fundamental conceit that emerged was the assumption that once a project was completed, the public was offered a fait accompli that they would naturally buy or use. The underlying theme of this position seemed to be an arrogant assertion: *Don't tell us what you need. Trust us to know what you want.* The result of such attitudes was predictable: customers went increasingly to companies whose projects and products reflected a concern for the customer, as illustrated by the phenomenal success of the Ford Taurus.

As a result of the new rules governing global business in the 1980s and 90s, project management must adopt a new standard by which future success will be measured—the so-called quadruple constraint. The additional feature of the quadruple constraint requires us to include *customer satisfaction* as one of the pillars of project success. Client satisfaction refers to the idea that a project is only as successful as it satisfies the needs of its intended user. This addition has tremendous implications for the way we manage projects and the manner in which the success or failure of both past and future projects will be assessed. With the inclusion of customer satisfaction as a fourth constraint, project managers must now devote additional time and attention to maintaining close ties with and satisfying the demands of external clients.

Among the implications of this new quadruple constraint is its effect on what were once viewed as traditional project management roles. Concern for the client, due to its extreme importance, necessitates that project managers adopt an outward focus to their efforts. In effect, they must now become not only managers of project activities, but sales representatives for the company to the client base. The product they have to sell is their project. Therefore, if they are to facilitate acceptance of the project and hence, its success, they have to learn how to engage in these marketing duties effectively.

The Assessment of Success over Time

Organizations often fail to make adequate allowances for the impact of time on a project's viability when assessing the performance of their project development. To illustrate, consider a real-life situation in which a company was determining the success of a recently completed hardware computer development project. Based on internal cost accounting data, the project looked good: It had come in on time and only slightly over budget. Further, the hardware performed as it was intended to. As a result, the project manager was given a performance bonus and reassignment for a job "well done." Unfortunately, the story does not end there. The project, while internally efficient, was a disaster in the marketplace from its first introduction. The technology that the company had assumed would be adequate, turned out to be so user unfriendly that the product was withdrawn within nine months.

This story illustrates a number of the problems we face in studying the causes of failure. First, it was clear that on the surface, this project was not seen as a failure at all; in fact, just the opposite was the case. The second problem had to do with the incomplete picture of project expectations that top management painted. Obviously, client satisfaction was never held up as a concern of the project manager who devoted his time to the measures that *did* matter for his performance appraisal: schedule, budget, and performance. Finally, the story demonstrates a subtler third point: it is important, in the absence of full information, not to assume that a project is a success or failure too early in its life, before the final "returns" have had an opportunity to come in.

The above conclusion does not only mean to suggest that many projects deemed successes are, in fact, failures. The reverse is also true: Many projects that appear to be instant failures may actually prove to be long-term successes. One example that comes immediately to mind is the well-known English Channel tunnel project, known simply as the Eurotunnel, or "Chunnel." Opening in 1994, nearly eighteen months behind schedule (late enough to miss the vital summer traffic season), the Chunnel project was originally budgeted for 2.3 billion pounds. The final bill, at 15 billion pounds, is well above six times the initial projection. From an internal auditing perspective, the Chunnel represents a financial nightmare, particularly in light of recent evidence suggesting that there is a potential for default on the bond financing made by the initial investors in the venture. Nevertheless, looking at the project in regard to its long-term potential, one must admit that its contribution to the human condition may be significant; in effect, leaving the judgement of project "success" or "failure" in the hands of future generations. We will discuss the Eurotunnel more fully as a separate case, suggesting that while a present day financial disaster, it is important not to write the tunnel off as a complete failure too soon.

This case illustrates the importance of balancing immediate assessment against long-term project viability. Clearly, there are definite benefits involved in waiting until after the project has been completed and is introduced to its intended clients before assessing the success and impact of the system. On the other hand, we must be careful not to wait too long to determine a project's external impact since the possibility exists that other organizational or external environmental factors may influence the organization's operations to the point where we are unable to determine the relative impact of the project in the marketplace.

For example, in the field of computer hardware development, product launch windows are often calculated in terms of months; either a computer is introduced during a favorable launch period or that company will be so far behind the technology and marketing curve that they may just as well not bother with the project at all. In this hyper-shortened timeframe, a "successful" project is assessed within months of its introduction. In contrast, the viability of Denver's new International Airport may not become clear until years into the future. In the meantime, a number of factors external to the project itself (e.g., the price of jet fuel, profitability of major airlines, or demographic patterns within the city itself) will serve to either elevate the airport project to dramatic success or condemn it to white elephant status.

Figure 3.1 illustrates the difficulty faced by project managers and their organizations (Pinto and Slevin, 1988). This figure shows a simple time line, demonstrating the point at which various aspects of project success can be evaluated. At the earlier stages in the implementation process, the typical assessments of success

FIGURE 3.1 Assessment of Project Impact Over Time

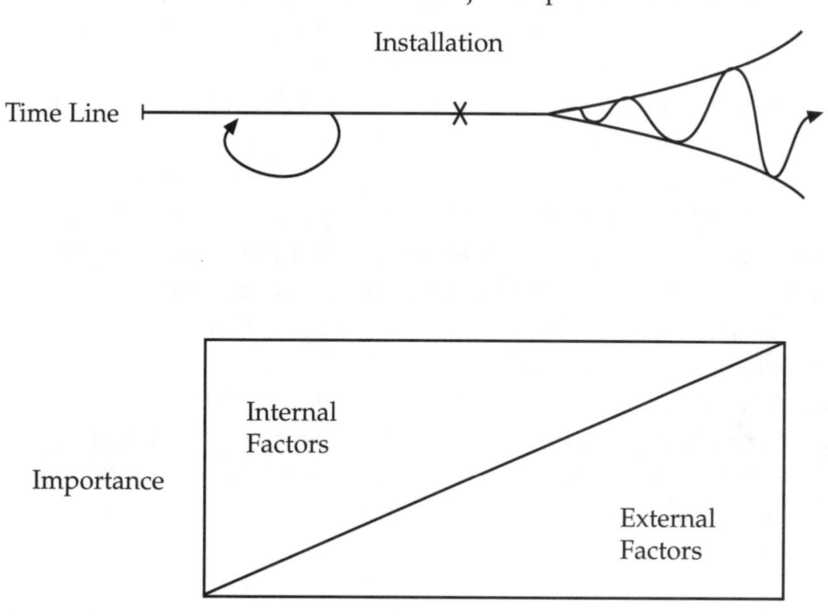

tend to revolve around those previously mentioned internal organizational factors such as adherence to budget and schedule, performance capabilities, project team cooperation and productivity, and project team-parent organization relationships. The loop in the system represents the instance or instances when periodic assessments of these internal criteria are made. If major discrepancies exist in any of these criteria, the project development effort may be extended, reconsidered, rebudgeted, or even scrapped. About midpoint in the time line is the point of project transfer. At this stage, a number of projections regarding the project's success are being made. The project is examined in terms of its technical capabilities. Have prototype results been satisfactory? Where appropriate, are the beta test site findings promising? While performance capabilities will become and remain very important, some other traditional measures of internal success become less so. For example, by this point, marketing and/or project transfer and installation schedules have been completed. The scheduling that remains pertains largely to client training and, if necessary, bringing additional features online. Further, the majority of the budget has already been expended at this stage. On the other hand, the issue of client acceptance and use is continuing to become of greater and greater importance now that the project has been completed, put in the marketplace, or handed over to its prospective users, and is up and running.

As the time line continues on to the right, it becomes progressively more wavy, symbolizing the increasing difficulty in accurately measuring project success based on concepts such as client impact and their perceived value of the project. Because the project has been completed, and for all intents and purposes is beyond the control of the developing organization, its fate resides with the company's ability to have accurately gauged the public's desire for the project. Consequently, external issues such as the state of the economy and changes in customer needs and tastes will continually impact the completed project. The project may initially be hailed as an important breakthrough, but down the road it may suddenly become both obsolete and an expensive, embarrassing failure (the Ford Edsel). On the other hand, some projects that were regarded early on as a waste of money or worse, may now be viewed as being very important to the organization and be considered successes.

Figure 3.1 suggests that while periodic assessments of the current state of the completed project are important, an accurate determination of the ultimate success or failure of a project is equally important. The difficulty lies in attempting to find a suitable reference point, at which time the project has been completed, transferred to the clients, and is having some initial impact in the marketplace. One of the important benefits of using such a "post-completion" project assessment is that it drives home the point to many project managers that their duty to the project does not end when it is "completed" from an internal evaluation perspective. In fact, in most instances, there is still a lot of hard work ahead. Project managers and their organizations can foreshorten some of their time with post-completion client involvement depending upon the degree to which they and other relevant project team members consulted with clients at earlier stages in the development process. However, it is absolutely central to our thesis in writing this book that an accurate determination of the ultimate success or failure of any project largely rests with the key external factors: the client and their use and satisfaction with the project. The main point underlying this position has to do with the organization's desire to fulfill the client's needs: matching the project to the client rather than attempting to alter the client's perception of their needs to fit the completed project. It behooves project managers to spend sufficient time involved in post-introduction activities as is necessary to better assess the "true" success or failure of the project.

POINTS TO REMEMBER

The discussion in this chapter has highlighted a number of salient issues that managers need to consider when attempting any sort of system implementation assessment. These points are enumerated below:

1) We have to balance cost consciousness with project performance.

There is sometimes a strong temptation to come down on either side of the cost/performance argument, usually with questionable results. For example, organizations that build projects at any cost because they are enamored of technology for its own sake inevitably run into trouble with project financing. We cannot continue to pour money down the gaping hole of a project run out of control. On the other hand, when project development is left in the hands of the accountant (i.e., project decisions are ultimately made almost solely on the merits of cost), we run the equally unpalatable risk of stifling crucial creativity and long-term innovation, all in the name of immediate capital constraints.

Clearly, a middle ground is needed—somewhere between the excessive controls of cost accounting and the anything-goes mentality of those who would tinker with technology just for its own sake. This conclusion forces management to adopt some decision rules for new product development that leave the project team as free as possible to ensure that performance is top-notch while at the same time ensuring that the team does more than simply pay lip-service to cost constraints.

To illustrate, consider the true case of a large, international corporation whose forestry division worked to develop a state-of-the-art loader for timberland clearing. Because the organization was dominated by its engineering function, the project continued to run high deficits against the original budget as the engineers worked to get the technology right. Unfortunately, while the project was introduced (moderately late and over budget) as a technological masterpiece, it is now showing the telltale warning signs of overdevelopment. For example, one of the primary markets for the loader is in the forests of Indonesia. Unfortunately, the loader, equipped with such state-of-the-art gadgetry as LCD monitors and other computer readouts, is unsuitable for the rugged treatment it would receive in the Indonesian forests. Often situated hundreds of miles from the nearest servicing center, the forestry industry needs equipment that is rugged and basic, rather than leading-edge and high-tech. As one engineer ruefully admitted, "They needed a Chevy and we built them a Rolls Royce." Hence, a technological marvel in the hands of the developing engineers runs the risk of mediocre sales in the marketplace.

2) Do not be too hasty to judge a project as a success or a failure.

It is the rare project that is correctly judged to be a success or failure immediately following its introduction. The cost accountant may endorse a project's development process as being within budget and schedule constraints. Likewise, our organizations may have positive feelings about a project's viability in the marketplace, but only time will tell the true story. In reality, many current projects cannot do more than offer imprecise evidence of their future success at the time of their introduction. The obvious problem is that we must make a serious evaluation of the project manager's role based on a project's future expectations. To what set of criteria is it appropriate to hold project managers in evaluating their performance? Clearly, we have argued that rewarding a project manager on the basis of adherence to the internally focused triple constraint is inadequate and, in fact, likely to send the wrong message to those running current development projects. On the other hand, it may be highly unfair to hold project managers to a standard that is truly beyond their control (e.g., the future stream of profits for a project with a ten or twenty year market cycle).

Some middle ground is necessary between the equally misleading guidelines of short-term development and long-term profitability. Companies that are intent on making their evaluations on as clear a set of data as possible must look for methods to evaluate client satisfaction, acceptance, and use of new project offerings in conjunction with the traditional measures of performance, schedule, and budget. Project managers need to be made aware of the standards that the organization holds with regard to its evaluation criteria so that they are willing and even eager to make the investment in client consultation and project marketing that are so important to ultimate project success.

3) Any project is only as good as it is used.

One of the most important points that past research and experience has taught us is that the client is the ultimate determinant of successful project implementation. This lesson, seemingly so fundamental, is one that nevertheless must be continually relearned by many organizations, both large and small. As long as our gaze remains firmly fixed on the internal criteria for project success, the so-called triple constraint, we will never achieve the full measure of the project success.

We live in a culture that is fascinated with the latest technological advances. Organizations that are dominated by their engineering departments are often particularly vulnerable to this bias. The prevailing assumption that one has only to build the latest design with the "hottest" component properties to ensure client acceptance, use, and subsequent project success is a conceit our companies can no longer afford to reinforce. When marketing and customer service are relegated to supporting cast roles, we will simply recycle the same mistakes again and again. In our drive to innovate, there is a very real danger that we will continue to pursue technology for technology's sake, rather than working to create projects that have practical and useful features. The oft-repeated statement, "Of course it will be used—it's state of the art!" reveals a high level of naiveté about how innovations are received and perceived by the average person. Project managers need to devote as much time to networking with potential clients as they do to ensuring that all the technical features of a new project are performing well.

4) Successful projects may require product *and* client modification.

Any new project must be "right" for the client base for whom it is targeted. Further, the "right" project refers to the importance of matching the needs and attitudes of the client base to the new product. Many clients acquire products that are underutilized because they were inappropriate for their needs. However, it is necessary that the reader understand that an "inappropriate" project is often not the direct result of technical difficulties or performance characteristic flaws. Sometimes a project will be perceived as inappropriate because it does not conform to the attitudes and value systems of the target population. In other words, the cultural ambiance of a firm or customer group must be taken into consideration when determining their needs.

We suggest that the process of developing greater acceptance of innovative projects involves a process of mutual adaptation between the project itself and the customer. Significant prework is required from the project manager and team members as they scan the client and objectively assess attitudes and needs regarding a project. If the team determines that it is not feasible to introduce a project within the current organizational or environmental context, they need to begin formulating plans for how to create a more supportive environment. That process may require either modifying the project to suit the technological needs of the client, engaging in large-scale training or education programs within the external environment to create an atmosphere of acceptance, or both. Unless project managers and their teams work to address potential problems of project acceptance, their highly developed and technically sophisticated innovations are likely to fail without having been given a sufficient chance to succeed.

CONCLUSIONS

This chapter has sought to address some of the confusions that result from our attempts to accurately assess projects as either successes or failures, arguing that there are a number of criteria that can cause confusion in our ability to first identify failures before learning from them. If failure is to serve as a valuable learning mechanism for future activities, organizations must be straightforward regarding their failures, just as much as they trumpet successes. When we are continually unwilling to learn from past mistakes, first by refusing to recognize failures, and then glossing over their impact, we simply perpetuate a recurring cycle of attempt—failure—attempt—failure.

Failure is first and foremost a client-based phenomenon. It is the client who is the ultimate arbiter of successful project implementation. Having said this, we should also note that an over-emphasis on client concerns, while willingly sacrificing internal constraints such as budgets, schedules, and performance, is not the answer either. A balance is required: an a priori decision rule that allows us to correctly prioritize our activities while at the same time ensuring that the project is not done in by a factor we could control but chose not to address. If such a balance is achieved, it will go far toward creating an atmosphere in which project priorities are well-understood and serve as guideposts for our efforts, reducing the "manageable" reasons for projects to fail.

References

Cringley, R.X. 1994. "How to forfeit millions for nothing," *Forbes*, vol. August 29, pp. 61–64.

Pinto, J.K. and D.P. Slevin. 1988. "Project success: Definitions and measurement techniques," *Project Management Journal*, vol. XIX, no. 1, pp. 67–72.

4

Project Management = Cost Control

CONTINUING THE DISCUSSION from the previous chapter, we examine some of the relevant research on project management. The title of this chapter represents by far the shortest, simplest, and perhaps the most effective definition of that all-embracing term, project management.

PROJECT MANAGEMENT—SOME GEMS

At the outset, let us share with our readers, some additional "gems" in project management. In keeping with our theme of exploring the lessons from project failure, consider the practical value of these dicta, many gleaned from our own experience with projects:

- Rarely anywhere does a new project, especially a high-tech undertaking, come on line on time, on budget, and up to expected performance.
- A project 95 percent completed is meaningless unless qualified by a statement such as, "half as much work still remains to be done."
- When an overrun becomes serious, try to seek the causes first. Next rework the project from the ground up, and if necessary, be prepared to abandon the project. Minor cost cutting to buy time is counterproductive.

The last dictum is based on a common sense that, unfortunately, seldom seems to prevail, judging by the number of totally nonviable projects which get "completed" nonetheless. This book and an earlier one (Kharbanda and Stallworthy, 1984) are proof enough of this. Why is it so often almost impossible to kill a project that is going badly? This question has spawned a number of answers to attempt to explain the irrational behavior behind the escalation of commitment to bad decisions. David Davis (1985), a British manager, offers his perspective on why bad projects usually will not be allowed to die. His reasons include:

- Obsessive enthusiasm by project managers
- Fascination with novelty and technology
- Good, old-fashioned need to achieve

What does he suggest can be done when an overrun does occur? He starts with some seemingly innocent but difficult questions, which include:

- Does the project involve pioneering technology; is it an R&D project?
- Has your company and/or project manager handled a similar project before?

- Is the project a mega one, bigger than the company has ever attempted?
- Was this project costed even before the design was completed?

A surprisingly large number of projects are approved long before design is completed. These include nuclear power plants, especially in the first generation, synthetic fuel plants creating oil from coal, and space exploration projects. Included in the last category is the high priority and prestigious project of landing a man on the moon. In the early 1960s, the (then) Soviet Union was ahead of the U.S. in terms of space technology. An unmanned space ship with a dog on board had orbited the earth as part of their preliminary attempts at landing a Soviet cosmonaut on the moon. This state prompted President Kennedy to commit the enormous resources of the U.S. to developing our own moon landing project. A very daring and most ambitious project, NASA operated with minimal budget oversight and virtually no limit on the expenditure. The results of the story are well-known: Scientists and engineers worked frantically to achieve President Kennedy's vision and ultimately succeeded in July 1969.

Davis concludes that it takes a brave manager to recommend the abandonment of his/her project after, say, a multimillion dollar commitment. Such an attitude can be encouraged by creating an environment to reward honesty and courage. This sort of atmosphere is not common, yet there are some notable exceptions. General Electric Company, for example, has lived up to CEO Jack Welsh's philosophy regarding new project ventures. One of his abiding beliefs is to create an attitude in which "failure" is not cause for termination or other career-ending reassignments. As he has suggested, it is vital to maintain an attitude of risk and excitement for innovation. The worst move that a company can make is to punish failure because it will serve to stifle risk. Better, he suggests, to reward risk, even if failure results, because one solid success can more than offset the expense of nine failed ventures.

Certainly, it is nice to create an atmosphere that encourages projects to succeed, but project managers must also know when to allow them to fail. It is a hard decision, but more often than not, it turns out to be the right one.

HIGH-TECH AND PIONEER PROJECTS

The formal project approval procedure works very well when the project can be properly defined, but high-tech and pioneer projects present a very different problem. Such projects are usually sanctioned well before the research and development work has been completed. The situation that confronts the project manager in such cases has been graphically described by Gaddis (1989: p. 146):

> *"[It is like] flying blind assisted by a relatively unproven set of instruments. His (project manager) experience, judgment, and faith must carry him through until early test results become available: from this first feedback he can modify the design approach . . . Meanwhile, during these periods of blind flying, he may be forced to make long-term decisions which commit substantial funds."*

Typical of the problems encountered in such situations is the actual case story of a project developed by a large corporation on the basis of a contract with the U.S. Department of Defense. The project was intended to develop a small, liquid fuel missile for a military objective. This technology called for a new, high-capacity pump. Rather than work to modify an existing pump, it was felt that there was sufficient time built into the project schedule for developing and proving such a pump, which was therefore subcontracted to a reputable pump vendor. The major contractor set up a mini-project for this purpose. As an added safety measure, a second pump vendor was also charged with the same task, but with a somewhat different design approach. In due course and within the time target, one of the two pump suppliers completed a prototype, placed it in a test loop, and proved its performance. But the designers were not satisfied. In the past, successful prototypes had been known to fail once brought into production, so a series of changes were introduced. This fine-tuning brought about a technological crisis in the life of the overall project, and enormous efforts had to be made to ensure that the prototype missile was ready for test with only a minor time overrun. Although no mention was made of final cost, it is obvious that the development and tinkering process caused significant overruns.

This story offers two morals, each important in its own right. First, there is often an inverse relationship between engineering domination within any firm and its overall capacity to bring products to market in a short time span. This is not to make the blanket argument that engineers cannot meet project schedules; rather, it is to suggest that technological challenge (which most engineers love!) presents a set of time frames that are usually not conducive to rapid resolution. Nor, for that matter, are engineers likely to offer the "quick and dirty" solution in order to save time if they can instead seek the optimal, elegant approach. Engineers seek technical challenge and time-consuming, optimal solutions—concepts that do not often fit within the cost accountant's initial project audits or budgeting estimates. The second moral follows closely from the first: It is extremely difficult to offer anything more than a best guess as to time frames with most pioneer projects. The organization is exploring uncharted water, rendering time estimates at best tricky, and at worst, absurdly optimistic.

Indeed, this characteristic of pioneer projects has become known as the "Concorde syndrome," named for the infamous Concorde project discussed later in this book. Cost overruns from five to ten times the original estimate are common. This phenomenon has been studied in depth by the Rand Corporation, and seems to apply to a wide variety of pioneer projects (Merrow, Phillips, and Myers, 1981). These findings, which have been largely corroborated by a number of other studies, reported that 80 percent of the typical pioneer projects studied failed to achieve their main objectives. One banker's experience seems to confirm this very readily (Arnold, 1989). He suggests that:

- Most large projects encounter large problems;
- These can be financially destructive for the company;
- Risk is often minimized by the use of worst case analysis;
- In one out of five projects, managers regret their decisions;
- The treacherous business environment upsets project promises.

Those proposing to become involved in major projects need to remember that the bigger the project, the more money it will lose if it gets into trouble. No major capital project should be undertaken without a realistic worst case analysis. Problems with the project can be accentuated by the fact that when a company does get into trouble, its bankers may well suspend further cash in-flow, making a bad position worse and initiating a vicious chain reaction.

All new projects, especially pioneer projects, or those with a high-tech element, carry a substantial risk. Costs can escalate and operational ability may decline as a project develops. Poor cost estimate and improper budgeting increase the risks, and enormous cost and time overruns can result. The natural response is an arbitrary tightening of the financial constraints, but this can easily lead to false economies that further cripple the project (Davis, 1985).

THE KEY AREA: INITIAL ESTIMATING

Proper project management really starts with a proper cost estimate right at the project conceptualization stage. As project definition and design progress, the cost estimate is continually refined until the budget approval stage. This preapproval estimate is supposed to be the most accurate cost estimate. Yet, it is rather paradoxical and even axiomatic that final actual cost is closer to the first "guesstimate" than to the later, supposedly more refined, estimates. Though a seemingly odd rule of thumb, many of the experts in the field of cost estimating seem to agree on this issue. Why this phenomenon is the case is a matter of conjecture, although an interesting and rather unusual paper claims to clear some of the other mysteries in this field (Skitmore, Stradling, and Tuohy, 1994). They explore conceptual estimating expertise using an extensive mail survey among U.K. quantity surveyors. Some of the significant conclusions follow:

- "Expertise" is not a universal phenomenon, but rather very project-specific.
- Different building types demand different approaches.
- Special attention is required for complexity of the project.
- The easiest projects to estimate are industrial factories and residential houses.
- Office construction projects are the hardest to estimate, perhaps due to the variety of designs/options.
- The most crucial attributes of good estimators are knowledge and care.
- There is no evidence of any individual having mystical inborn talent for cost-estimating.
- Good estimators have exactly the same attributes as good gamblers: they weigh each decision against possible up- and down-side risks and behave accordingly, and they research selectively and thoroughly.

Project Monitoring and Evaluation

In many respects, the 1990s are a decade of challenges and opportunities for project monitoring and evaluation (M&E). The challenge lies in widening M&E responsibility to assess both implementation and planning "failures." Monitoring and evaluation represent the core of any system whereby project management can be assessed in "real time." In other words, as a control mechanism, M&E offers unsurpassed techniques if properly used. It is highly desirable that M&E survey results are made available as a matter of course to all concerned, and confidentiality be kept to an absolute minimum, if we are to learn from the process.

There are three alternative methods for employing M&E methodologies to control projects: Feedback, Concurrent, and Feedforward. Feedback forms of M&E are the most elementary and the most widely used. Essentially, they consist of any mechanism which allows us to assess the status of our project following the occurrence of significant events. For example, exception reports give project managers a view of the project status by reporting on what it has failed to do. On a monthly or quarterly basis, a set of exceptions are generated to allow project managers to see what areas are deficient. Note that although of some use as a future learning tool, exception reports indicate what has already happened, not what is occurring or is likely to occur in the future. Consequently, as a monitoring device, feedback information is the equivalent of closing the barn door after the horse has bolted. At best, we can hope that the information conveyed is helpful for future activities or even different projects because it does little to help the one currently under way.

Concurrent monitoring requires that the project team have access to sophisticated M&E techniques that can provide real time data. The underlying theme of concurrent project control is to provide the information in a timely manner in order that this precise information can be "actionable;" that is, it is current enough that it can form the basis for immediate corrective action. Sophisticated project tracking software often serves as a form of concurrent M&E, provided it is updated on a regular basis.

The idea behind the final type of M&E, Feedforward, is that present actions have future implications. Suppose, for example, that a small-to-medium sized company wins a major contract with a large manufacturing organization. The contract is expected to start in three months. The subcontractor's first requirement is to make a realistic assessment of the current facilities, capital equipment, and personnel. If it is determined that they do not have the facilities on site to create a large project team necessary to complete the contract, they will look for additional office space now, rather than waiting three months until the clock has begun on the contract. This is an example of the sort of proactive, feedforward M&E that can be most useful to project organizations. It requires them to look forward and make a series of "What if?" judgments and operate in an anticipatory, rather than reactive mode.

When M&E are used in the above manner, they offer project managers an important exercise in strategy making. Monitoring and evaluation become more than a simple afterthought check on how we did; rather, they allow companies to determine where they are going. From the perspective of assessing the causes of project failure, it is clear that one look forward is easily worth two looks back. This point was also born out in large-scale project management research that looked at the causes of project failure (Pinto and Mantel, 1990). The findings of this study demonstrate most dramatically that the number one cause of project failure was the inability or unwillingness of the project manager and team to engage in the "What if?" trouble-shooting guesswork before problems came to the surface. Indeed, they found that the vast majority of projects whose schedules spun out of control were those that ignored any sort of forward-looking activities, preferring instead to concentrate on the present status of the project. Unfortunately, present day complacency is poor preparation for future shocks.

TEAMS CAN ACHIEVE WONDERS

In a two-part article by a team of owner and engineering-construction contractors (Boutte, Matuzak, Pike, and McWilliams, 1994), we are literally taken behind the scenes of a specific project, namely the construction of a coking unit. The first part deals with the activities which facilitated construction, thereby minimizing project duration. The second part discusses integrated control systems for project and system monitoring, as well as system turnover strategies and procedures. It concludes with the lessons learned from execution of the project, which is indeed the spirit of our present effort.

This project was a resounding success, particularly when measured against initial cost estimates. Probably the single biggest contributing factor to cost containment was the project team's ability to keep a handle on initial materials usage and activity schedules. The project schedule was optimized, thanks to intensive teamwork. Each member of the project team was directly challenged to find ways in his or her particular area of expertise, to trim excess "fat" from the budget. A summary of the payoff shows the benefits for the team:

Original project cost:	$430 million
Original completion time:	36 months
Savings as a result of:	
Completion in 31.5 months	18 million
Surplus material 3% instead of 5%	1.1
Constructability 3% instead of 1%	12.9
Rework 2% instead of normal 5%	0.9
Supplier quality improvement	0.9
Startup 19 days instead of 30	1.3
Safety 3.8 against 12.9 average	3.0
Total Savings:	38.1 million

How were these savings achieved? Briefly, through:

- Emphasis on total project team participation both initially and throughout the process
- Full-time key personnel seconded to the team in early project planning
- An Information system developed to consolidate various project needs
- Commissioning/startup integrated into construction planning

- Smooth flow downstream: engineering, procurement, construction, startup

The savings were obvious, but more importantly, the key lessons that were taken away from this experience have gone far toward improving project management throughout the organization. Some of the important lessons that were learned included:

1) The effective use of an outside facilitator—An objective expert was retained to trouble-shoot the development process, offering suggestions for improvement and critiquing the systems in place.

2) Effective expectations matrix—At the beginning of the project, all key suppliers and subcontractors were assembled and given a comprehensive overview of the development sequence, paying particular attention to their interdependencies.

3) Early and simultaneous mobilization of engineers and constructors—Rather than relying on a linear flow model of responsibilities, the project collaboratively involved the principal actors at the beginning of the process. Consequently, rework and wastage were drastically reduced due to an enhanced understanding of the capabilities and expectations of each key player.

4) Active community involvement—The project team took a proactive posture in anticipating any potential negative reactions to the construction from potential "intervenor groups" in the external environment. As a result, rather than waiting for bad publicity and start-up delays, the project team was able to communicate a sense of partnership with these external groups and smooth the transition to private operation.

A Fast-Track Schedule Helps . . . Sometimes

Project management has turned out to be a natural discipline, lending itself to the total quality movement. A fast-track schedule with a built-in quality dimension can go a long way to facilitate the success of any project, particularly in terms of its completion on time and within budget. Typically, a knowledgeable owner will ask the engineering/construction contractor to complete a project in the shortest possible time. When the contractor submits the appropriate schedule acceptable to the owner, it is likely to be accompanied with conditions, such as:

- no changes
- minimum interference
- limited review and approvals
- full use of contractor's standard specifications/designs
- worldwide material sourcing, vendor preselection, limited bidding
- full control of construction labor (Kerridge, 1990)

Admittedly, the obvious drawback with these points is that they represent the perfect world, one in which all specifications are immediately frozen, client interference during development is minimized, and full control of labor sources is maintained. Nevertheless, when contractors have a degree of freedom that is similar to the terms stated, they have the option to begin planning in terms of overall coordination of activities. This is truly a fast-track schedule. And this can only be achieved through constant, unremitting battles fought and won every day in order to hold to the milestone schedule established beforehand. Of course, the master plan must have a built-in flexibility to provide for unforeseen events, knowing that they are the rule rather than exception. The maintenance of the master schedule requires continuous and unremitting review and monitoring of all activities, at the headquarters, project office, and in the field.

Fast-track schedules have their share of pitfalls as well. Project time can indeed be compressed, but usually at a cost. Consider that, as a general rule of thumb, compressing project time by one percent means an increase in the project cost by one to two percent or even more (Graves, 1989). The time-cost tradeoff curve is the usual asymptotic type (See Figure 4.1).When time is equal to zero, costs are infinite and vice versa. A detailed knowledge of such cost/time trade-offs for a firm vis-a-vis its competitors can be a vital factor in formulating competitive strategy. While the article cited above relates to R&D projects, we venture to suggest that the relationship will broadly hold for projects in general. And the percentage increase in project cost for a one percent reduction in project duration for various types of projects is stated to be 1.75% for hardware projects and 1.9-2.0% for software projects.

To illustrate this point, consider a hypothetical example of two competitors trying to develop the same product with the following parameters:

> Project duration: 5 years
> Development cost: $100 million

FIGURE 4.1 Time-Cost Tradeoff Curve

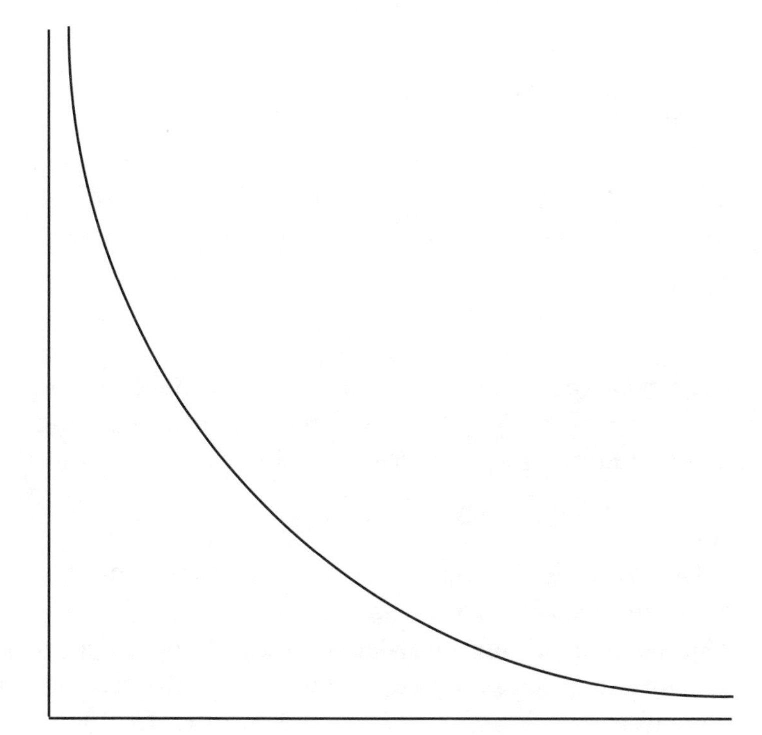

Time

Cost

The additional cost for one company to complete the project 10 percent ahead of schedule will be about $20 million. However, the other firm with less experience or fewer resources is likely to incur more than $20 million in additional costs. These costs are included in learning curve development and the extra time necessary to complete the project due to lagging schedules brought on by insufficient resources. Consequently, there is an interesting dynamic at work: crashing the project activities to finish early can add to the project's final budget. On the other hand, a meandering schedule and habitual delays can also add tremendously to the final cost. The project manager must always keep in mind the time-cost tradeoff before deciding to "crash" a project.

Overall Project Cost Control

While the contractor or owner (depending upon the situation) has the largest stake in the project and ultimate financial responsibility for it, he has to delegate some of these responsibilities. But he still has a very significant role to play, since it is his project and his money! There are two key functions which must remain under the direct control of the contractor, namely finance and cost control. These activities will continue alongside those of the contractor and his personnel, but their work will continue even after the project has been commissioned and the contractor has left. Several books have been specifically devoted to the guidance and instruction of such promoters (e.g., Stallworthy and Kharbanda, 1983; Kharbanda, Stallworthy, and Williams, 1987). What has to be realized is that the expenditure record has no real value for project cost control.

Cost recording is not cost control. Cost recording is usually an accounting function and comparatively easy, whereas cost control requires engineering knowledge and judgment and is very time-consuming. A sound evaluation of project progress demands time and effort on a continuous basis, not at infrequent intervals. In the contractual relationship under consideration, the detailed work of assessment will be carried out by personnel under the direction of the managing contractor, but the project manager must ensure that work is done properly, thus not relying entirely on the expenditure record. There are techniques available which enable sound assessments to be made to an appropriate level of accuracy with the minimum of effort (Kharbanda, Stallworthy, and Williams, 1987).

The commitment record is another detailed cost record that is a powerful tool for detailed cost control. The commitment record has to be complete and comprehensive if it is to serve its purpose within the cost control function. The commitment on some orders and contracts has to be estimated, and rarely does actual expenditure precisely equal the commitment recorded. This means that as the project approaches its end, the commitment record needs to be carefully scrutinized for validity. Outstanding variances are not always deleted as orders are finalized, contracts closed, and claims settled.

Project cost control is a seemingly tedious undertaking that lacks the glamour of the technical side of project development. Nevertheless, as this chapter has pointed out, it is precisely in keeping these detailed records that project managers stand in the best position to keep a tight rein on the project and ultimately go far toward ensuring its success. Runaway budgets and schedules are two of the surest signs of projects that have either failed or are well on the road to disaster. True, R&D projects present special challenges for project monitoring in their trek across unexplored territory. There is, however, a difference between granting engineering leeway for discovery and writing the equivalent of a blank check. Readers who are project managers must keep clear the point that technologists do not necessarily operate under the same set of constraints and time frames that other members of the project team adhere to. Consequently, there is a very real juggling act that project managers need to perform to keep a proper balance between technical demands and schedule and budget requirements.

As we stated at the beginning of this chapter, project management is cost containment, the often laborious process of tracking and staying on top of various expenditures and schedule decisions. Each decision implies a trade-off: In order to get the technology perfect, I am sacrificing time. In order to crash a project, I am giving up budget money. Each decision, right or wrong in hindsight, must be made with a clear head, based on the best available information. Most importantly, these decisions must be made with an eye toward the prize. After all, it is your project out there!

References

Arnold, J. H. III. 1989. "Assessing capital risk—you can't be too conservative," in *Managing Projects and Programs, Part III, The Project Manager.* Cambridge, MA: Harvard Business Review Books.

Augustine, N.R. 1994. "L.A.Engineer," *The Bridges (National Academy of Engineers)* vol. 2 (3), pp. 27–29.

Boutte, F D, Matuzak, P.J., Pike, W.G. and McWilliams, D.N. 1994. "Optimize Project Execution: Part 1," *Hydrocarbon Processing,* vol. 73 (1), pp. 116–121.

Boutte, F D, Matuzak, P.J., Pike, W.G. and McWilliams, D.N. 1994. "Optimize Project Execution: Part 2," *Hydrocarbon Processing,* vol. 73 (2), pp. 109–114.

Cameron, J. 1993. "The challenge for monitoring and evaluation in the 1990's," *Project Appraisal,* vol. 8, June pp. 91–96.

Davis, D. 1985. "New Projects: Beware of False Economies," *Harvard Business Review,* vol. 63 (2), pp. 95–101.

Gaddis, P.O. 1989. "Managing projects and programs," *The Project Manager,* Chapter 1, Part III. Cambridge, MA: Harvard Business Review Books, pp. 145–162.

Graves, S. 1989. "Why Costs Increase When Projects Accelerate," *Research-Technology Management,* vol. 32 (2), pp. 16–18.

Kerridge, A.G. 1990. "Use a Fast-Track Schedule," *Hydrocarbon Processing,* vol. 69 (11), pp. 75–82.

Kharbanda, O.P. and Stallworthy, E.A. 1984. *How To Learn From Project Disasters—True-Life Stories with a Moral for Management.* Aldershot (UK): Gower.

Kharbanda, O.P., Stallworthy, E.A. and Williams, L.F. 1987. *Project Cost Control in Action, 2nd Edition.* Aldershot, UK: Gower.

Lawrence, A. 1991. "Nuts and Bolts of the Project," *Management Today,* January, pp. 84–85.

Merrow, E., Phillips K., Myers, C. 1981. *Understanding cost growth and performance shortfalls in pioneer process plants.* Santa Barbara, CA: Rand Corporation.

Pellegrinelli, S. and Bowman. C. 1994. "Implementing Strategy Through Projects," *Long Range Planning,* vol. 27 (4), pp. 125–132.

Pinto, J.K. and S.J. Mantel, Jr. 1990. "The Causes of Project Failure," *IEEE Transactions on Engineering Management,* EM-37, no. 4, pp. 269–276.

Skitmore, R. M., Stradling, S.G., and Tuohy, A.P. 1994. "Human Effects In Early Stage Contract Price Forecasting," *IEEE Transactions on Engineering Management,* vol. 41 (1), pp. 29–39.

Stallworthy, E.A. and Kharbanda, O.P. 1983. *Total Project Management,* Aldershot, UK: Gower.

5

Early Warning Signals

IN THIS CHAPTER, we wish to make note of some of the best early signals for project managers when their projects are in danger of going off-course. Paying attention to these early signs is one of the most vital, yet most neglected functions in project management. Why? It is usually due to management's inability to refocus from the micro approach that consumes much of the project manager's day, to a strategic, overall macro focus on the direction the project is taking. One manager explained this process to us: left unaddressed, the effect is equivalent to rearranging deck chairs on the Titanic. Even though the consuming minutia may be well-performed, the project as a whole is sailing slowly and purposefully into the midst of disaster. We will demonstrate that early action (and it can never be too early!) on the basis of little bits of information and trends is often far better and more effective than waiting until complete information is at hand. And the corrective action can be at any stage of the entire process right from the conception through design and execution to completion and commissioning.

We must point out at the outset that the term "early warning signals" is usually applied in project management texts to cover the period of project execution, taken as after the stages of project conceptualization, plan formulation, and even design. The project formulation and design functions are vital in terms of project cost control (an essential function of project management), but are unfortunately often ignored. For this reason, our present chapter deals only with these functions.

PLAN AND SUCCEED

There is no magic formula for project success. But if we were to offer one element that has far-reaching implications, our response would be in the form of a three-facet formula: plan, plan, and yes, *plan!* This point is not intended to sound facetious: Most, if not all major failures on projects can be traced to inadequate planning, inaccurate planning, and/or blind adherence to originally formulated plans regardless of how the environment has changed in the interim. We should note that the term "planning" here refers to both a set of processes (conceptual) as well as the physical plan (nuts and bolts). When planning is sufficiently addressed, project managers and their teams have devoted enough up-front time to create extensive documentation (e.g., Work Breakdown Structures and cost estimates) and realistic road maps (e.g., Critical Path networks and Gantt charts) to enable them to have a clear-eyed understanding of the roads they must travel in order to complete their projects successfully.

Many project disasters can be laid directly at the door of planning processes that are perceived to have gone wrong (Hall, 1982). Broadly, Hall notes that planning errors often comprise one of two distinct types: *Negative* refers to faulty or bad plans in themselves. These errors lead to physical results which are often substantially modified, reversed, or even abandoned (e.g., nuclear power plants and Iran Petrochemical Plant, in the current book are good examples of these errors) at a later date. The second sort of planning mistake is referred to as *Positive,* comprising acceptable action steps taken in the wrong direction. Positive failure includes examples of projects that were implemented despite severe criticism and opposition (e.g., The Sydney Opera House, Concorde Airplane, and the Chunnel). Naturally, all of these involved great public controversy, but Hall notes, quite correctly, that such judgments are purely subjective and are unlikely to lead to any form of unanimity.

If planning is so important to successful projects, why is it that we often find poor planning to be a major source of failure, principally the trigger in run-away project costs and schedules? The short answer is that project managers are under considerable pressure from top management to show some visible signs of project progress. In the absence of these demonstrated steps, upper management becomes increasingly worried and demanding of the project manager. The assumption is that if no "real" progress can be shown in the project's development, nothing must be getting done. Consequently, there is a strongly felt need for project managers to start big and continue making the kind of noise that convinces the people at the top that the project is moving along well.

Unfortunately, these sorts of pressures and the activity they induce are often counter-productive to project success because they force the project manager and team into the wrong mode (project execution) far too early in the process. One of the important messages we would convey to both project managers, and more importantly, top management, is to acknowledge the necessity of sufficient advanced planning prior to project activities. Indeed, we suspect that if an analysis were conducted at most corporations, they would discover that it is usually those projects that started off impressively, with minimal up-front planning and lots of frenetic activity, that ended in difficulty, either by requiring extensive remedial attention or abandonment.

Since the termination of the Second World War, the global economy has, with some bumps along the way, experienced an extended period of economic growth, initially in the developed world and more recently in the developing world. The disasters we describe here, as well as many others described and analyzed elsewhere, offer us some important reminders of the perils of rapid expansion, implying the often hidden side of progress: For every success, there is often a littered trail of miscues, errors, and even willful deceit. Progress does not come without a price, often ill-defined or time-delayed, but insistent nonetheless.

One of the recurring lessons that these case stories will illustrate is the tremendous impact that adequate and timely planning can make on projects both large and small. One of the most obvious but overlooked early warning signs of impending failure is the inadequacy of initial project planning.

LEARN TO FAIL, THEN LEARN FROM FAILURE

The above dictum may at first sight appear to be a contradiction in terms, but it contains an element of wisdom. Managers are so obsessed with success that it is the focus of most management literature. In contrast, our central message here is: Failure is an inevitable part of human endeavors, so accept it and try to turn this threat into an opportunity. How? By learning from failures, particularly the far cheaper course of learning from the failures of others. Hopefully, this process will give us the secret of how not to fail.

Consider the design process, the first step after the project is firmed up and approved. Real cost control starts at the drawing board. The more time, effort, and attention that is spent at this stage can pay enormous dividends downstream. Once the design is frozen, the scope for savings reduces drastically. However, the design process offers its own share of risks versus rewards. To illustrate, a distinguished academician, Henry Petroski, has made an in-depth study in the design area. His findings in the use of computers and design are intriguing. Petroski notes that in modern times, computers are increasingly seen as the "in" thing, representing a solution to all our problems. But all is not as it appears. Computer-Aided Design (CAD), while much in fashion, can, if we are not careful, very easily lead to computer-aided disaster! Why and how? The growing dependence on computers as a solution to all our problems leads to over-confidence and potentially, to disaster (Petroski, 1985).

Petroski's point is intriguing: As opposed to the current reliance on computers, the use of old-fashioned, manual calculations gives the engineer confidence and also the discretion to provide for an adequate safety factor. Such judgment is missing in the case of CAD, easily resulting in a design that borders on the dangerous. With easy and rapid computerized calculations, an engineer falls into the natural temptation of optimizing every part in order to save on weight and cost. Such "over" optimization was not typically attempted with manual calculations, resulting sometimes in overdesigns that were, nevertheless, inherently safer! It is suggested that even with CAD design, the designer should be able to oversee and even "overrule" the computer to ensure an adequately safe design (Petroski, 1989).

FREEZE DESIGNS AND NO CHANGES PLEASE!

It is one of our central themes that case studies of project failures form an invaluable source of information and if properly analyzed, can provide a real insight as to how human error can totally negate a seemingly successful project, especially in regard to the design process. Such studies can help identify some common sources of error in design, thereby leading to better understanding of the relationship between design and construction (Petroski, 1985b).

Although design "freeze" is most desirable, it assumes a static environment, which is often far from true. The intense effort to control project costs may prove futile if the several causes of unprofitability are not kept constantly under scrutiny. After all, the assumptions regarding the viability of a project are constantly changing as fresh information becomes available. A sudden shift in a project's direction dictated from the top can be very irksome, but reality forces us to acknowledge that project managers have to live and deal with the consequences.

Change is a constant—indeed, a mandate—in business today,

and change in design as in any other function is no exception. Design changes arise for a number of reasons. They can be the result of any recent failure of a similar structure or design, or due to technological breakthroughs that, if ignored, will render the project obsolete even before its introduction. Changes may also occur as the result of external pressures (usually from clients). Such changes, of course, are often for the better, but we must distinguish between changes which are absolutely unavoidable or others which are desirable. It is toward the latter that we need to focus our attention. The real cost of such a change is far more than what appears to be the nominal additional cost. As an example, consider a simple change of a valve size from the existing 2" to a more desirable 2 1/2". The apparent increase in cost may be just a few dollars, but the real cost increase may turn out to be a hundred times as much. How and why? This huge increase must necessarily take into account the changes consequent to the seemingly innocent "little" change. These include change in the pipe sizes connecting the valve, and change of drawings to conform to the plant "as built" with its respective additional time and cost. The numbers add up fast. Hence, the frequently heard plea from project teams, "No changes, please!" Let us, however, add, "To the extent possible."

Do Not Underestimate Estimating

After a project is firmed up and approved, design is the first major activity to be considered. Following this is the process of cost estimation. Ideally, design and costing should be dovetailed in a similar manner to the approach taken in Japan. Logically, the potential for cost control/savings is maximum at the early stage of freezing the design. Our normal procedure is design, then cost. Japan has reversed it to cost out the design. In effect, they start from the market, a possible selling price, then work back to the cost price in order to realize the project's required internal rate of return. As a result, the designers are given the challenging task of designing a product to that initially benchmarked cost (Kharbanda and Stallworthy, 1991). To give one well-known example, the Sony Walkman was one of the most successful creations of this seemingly "reverse" cost estimating sequence. To some readers, this process may sound a bit like "putting the cart before the horse." But it has turned out to be a highly effective move for the organizations that have adopted this pattern.

Cost estimation, however, does not always approach the level of being an exact science. In fact, it is subject to a number of vagaries. Early in his career (1960s), one of the authors was concerned with the fabrication of a nuclear reactor in India at a time when the local facilities were not geared for such sophisticated jobs. The story has been told elsewhere (Kharbanda and Stallworthy, 1983), but we bring you the highlights. Having had no experience in fabricating complex equipment with (almost) unheard of tolerances and precision, it was virtually impossible to create a reasonable advance estimate of the cost. The estimators did the best they could, then added a little more than normal margin, and quoted to the client.

Thereafter, we happened to attend a week-long international nuclear power conference that included stalwarts in this field from all over the world. About midweek, we were fortunate to come face-to-face with the chief engineer of the company that had supplied the first reactor to India, identical in design to the one we had recently bid. This was the chance of a lifetime to finally get the inside information on accurate cost estimating. In fact, the expert confessed that his company lost "their shirt" on the Indian reactor, a common experience of the first generation fabricators. Then in reply to our innocent question, "How do you estimate a nuclear reactor?" the expert answered with masterly confidence, "Do your normal cautious estimating, add more than normal margin and then after a short pause, double it!" In our ignorance, we skipped the last vital step, but this short, casual conversation proved most valuable. We were forewarned, we took it seriously, and got forearmed. It saved us several millions of dollars.

The importance of an estimate cannot be over emphasized. Project estimation is indeed a yardstick for project cost control. And if the yardstick is faulty, you start on the "wrong foot." It is in this context that we exhort you not to underestimate the estimate.

Our experience has shown us that estimating is undoubtedly an art as much as a skill.

Project Managers Can Do Wonders, but Beware of Murphy's Law!

A significant part of the art of successful project management lies in choosing the right people for the project you have in mind. But the crucial figure in the team must be the project manager, the one person who can indeed make or break a project. When a project overruns its budget, the project manager should first seek causes, then remedies. The project manager should be able to recognize when things are going wrong, and hopefully proceed to correct them before disaster comes. This is largely a question of leadership, so those qualities of a project manager are all-important. Yet even the best project managers can get into trouble, largely due to Murphy's Law.

Most of us are familiar with the tenets of the law: Anything that can possibly go wrong will. Murphy's own wording for the law, as enunciated long after the transducer incident, runs thus:

> *If there are two or more ways of doing something, and one of those ways can result in catastrophe, then someone will do it that way.*

Obviously, the underlying point is not that failure is inevitable. Rather, Murphy's law suggests that locking into certain assumptions about people and events can be highly dangerous. Too many things (both within the project manager's control and external to the project) can go wrong somewhere during the development cycle, causing all manner of headaches. Murphy's message is that wise project managers spend far more time planning for problems than anticipating success. Clearly, it is possible, and indeed likely, that in the midst of a highly complex project involving many subroutines and activities, someone, sometime, will do the wrong thing, but a lot can be done to prevent this from happening. This is precisely the role of the project manager and his team.

What is the lesson to be drawn from Murphy's Law? Is it not that human beings are fallible? This means that we must never make the assumption that people will invariably act reasonably, sensibly, or competently. This is a factor in human behavior that the project manager, and indeed all the members of the project team, must never forget. A good understanding of the quirks of human behavior will go a long way in ensuring that a project is completed successfully. By taking steps to troubleshoot and anticipate problems, the errors that would otherwise occur will be minimized. Indeed, one writer on this subject summed the situation up with the advice: "Try to change situations, not people." (Kletz, 1985) Another writer explains it more fully, suggesting (Belbin, 1981: p. 49):

> *The task is not to change people. People are perfectly alright the way they are. The task is not to motivate people. People are inherently self-starting. The task is to remove those things that demotivate them, to get them out of their way. Or, more precisely, to create those kinds of organizational structures that allow workers to get at problems and act in some independent ways so they can develop their skills solving problems related to their own jobs.*

So people will manage, if allowed to, but we will always have human error. Is there a way out? In our view, managerial competence is at the heart of the matter. It is mismanagement that gets projects into trouble: improve the quality of management and you go far toward avoiding disaster. It is for this reason that apart from analyzing a series of specific project failures, we develop significant aspects of good project management. There is literally tons of literature available on the techniques and guidelines that should ensure a successful project. But foremost among these is undoubtedly the transforming of the project team into a dedicated team of "managers," who right from the start are determined to agree upon a plan and implement it fully and successfully. That is the road to success.

CURE FOR ELEPHANTIASIS: PRE-APPROVAL AUDIT

The headline of a concise article proclaimed:

> *"Poor countries should make rich companies finance and run the projects they long to sell."*
>
> (*The Economist*, 1988)

Our heading above is derived from the title of *The Economist* article, inspired no doubt, by a host of "white elephant" projects, especially in developing countries. It is precisely such projects which cause countries like Brazil, the Philippines, and Mexico to reel under a tremendous burden of foreign debt. While similar projects do, to some extent, exist even in developed countries, they have the capital resources that make them better able to handle and/or survive such overruns. As an example, the former West Germany could take in stride a tremendously expensive but ultimately pointless petrochemical plant built on its North Sea coast. Likewise, Great Britain has the capacity to cope without much trouble with the cost of an international airport in Scotland, although it has never been used.

Projects such as these were built not because they were needed and justified for techno-economic reasons, but out of prestige or the desire to be self-reliant. As an example, it sometimes seems that every undeveloped country seeks to produce steel not only for their own requirements, but also as an export item in spite of the huge odds against success (e.g., raw material availability and economies of scale). Likewise, consider a country like India, in effect a sub-continent, in which most states seek to have their own steel manufacturing plant and petrochemical plant, almost as though these structures are a status symbol, even if used to less than half their capacity. Obviously such reasons do not make economic sense and have, indeed, led to a host of "white elephants." One way to avoid such pointless structures, as *The Economist* headline above has noted, is the so-called BOT (build-own-transfer) concept with its several variants.

Essentially, the BOT process requires the contracting (or constructing) organization to build the plant and operate it for a contracted period of time before turning it over to the customer. The benefits of such a policy for the local country are huge in that the constructors are willing to own the plant and thus, take all risk for its viability during the early stages of operation. As a result, it is far less likely that clients will have "white elephants" foisted on them by constructors who build them, get their money, and abruptly leave. When faced with making their own significant investment in the project, the company in question is forced in the most obvious way to "put their money where their mouth is."

Build-own-transfer procedures are useful but not widely practical, since the vast majority of commercial projects initiated by organizations are not of the magnitude or type that would warrant such an approach. Perhaps a far more effective way to prevent such disasters is to do thorough homework before sanctioning and proceeding with the project. This process involves the so-called preapproval audit, but unfortunately, this approach is much neglected. Among the pioneers of the preapproval audit for projects was the American company, Armco. Many companies use such audits in a very informal way, but the Armco approach was fully formalized, and thus serves as a very effective check, preventing the implementation of unprofitable projects. It makes capital budgeting orderly and efficient, since the profit center staff must complete a thorough analysis with documentation for top management acting as independent advisers (Lambrix and Singhvi, 1984).

How did Armco go about developing their preapproval audit? It started initially with a $50 million dollar project for one of its divisions, which proposed to buy continuous casting equipment back in 1978. The promoters of the project claimed a 22.6 percent discounted cash flow return over the machine's twenty-year life. Additional results have been equally positive, prompting continuation of the technique.

To get an independent analysis of the project, the management appointed a five-person audit team drawn from research, marketing, strategy, operations, and finance. The team questioned the division's assumption of 98 percent capacity utilization, since the actual average was likely to be closer to 74 percent. Giving the division's estimators the benefit of doubt, the team assumed 85 percent utilization, thus resulting in a discounted cash flow of 17.4 percent. This was further revised downward to 14.2 percent when the extra cost of pollution control equipment was added, a factor ignored by those making the original proposal. The project was in fact approved some two months later, but meanwhile, the division involved had prepared further supporting documentation, which proved extremely useful during the post-completion audit. This exercise became the model for the approval of all new projects and it proved to be an extremely valuable tool, ensuring that investment decisions were made more objectively and realistically.

From 1978 to 1983, some twenty large projects were handled in this way: five were rejected, five were approved with modifications, two were partially approved, and eight were approved as submitted. The high percentage (40 percent) of projects approved as submitted was believed to be due to the fact that the proposers knew that they would be subject to an independent audit, and therefore, did their preparatory work much more thoroughly. Several other companies have had those kinds of experiences and we would strongly recommend our readers to follow a similar course at project formulation and approval stage. That is really the time to deal with the early warning signals.

One of the best methods for dealing with problems is to avoid them in the first place. Clearly, it is impossible to anticipate all possible project warning signs. This chapter offers instead a select set of some of the more obvious or significant types. It is, however, an extremely important practice to devote enough time in the upfront period of any new project to conduct the sorts of analyses that we have suggested here. As both our own experience and research continually bear out, it is possible to avert a huge amount of down-stream problems by making a concerted effort to look for problems at the earliest possible stage.

References

"A cure for white elephantiasis," 1988. *The Economist,* Vol. 308, September 17, p. 20.

Belbin, R.M. 1981. *Management Teams—Why They Succeed or Fail,* Heinemann, London.

Kharbanda, O.P. and Stallworthy, E.A. 1984. *How To Learn From Project Disasters.* Aldershot (UK): Gower.

Kharbanda, O.P. and Stallworthy, E.A. 1991. "Let's learn from Japan," *Management Accounting,* Vol. 69, No. 3, pp. 26–33.

Kletz, T.A. 1985. *An Engineers's View of Human Error.* Rugby, UK: Institution of Chemical Engineers.

Lambrix, R.J. and Singhvi, S.S. 1984. "Preapproval audits of capital projects," *Harvard Business Review,* Vol. 62, March/April, pp. 12–14.

Petroski, H. 1985a. "Superbrain, superrisk," *Across the Board,* December, pp. 48–53.

Petroski, H. 1985b. *To Engineer is Human—The Role of Failure in Successful Design.* London: St. Martins Press.

Petroski, H. 1989. "Failure as a unifying theme in design," *Design Studies,* Vol. 10, No. 4, pp. 214–218.

6

Simple Is Beautiful

THERE IS REALLY no substitute for practical experience, yet a project manager has to temper this dictum with practical reality; we often cannot afford to learn the hard way. Not only is it time-consuming, but it is expensive. Of course, in every field of activity, there is always an element of risk. The real test of good project management is to try to reduce these risks to the barest minimum and/or acceptable level through meticulous planning, execution, control, and above all, motivation of the entire project personnel.

BUREAUCRACY CAN MESS THINGS UP

It seems a sad irony that our normal business systems place greater reward on ensuring that nothing goes wrong rather than ensuring that something goes right. Unfortunately, the alternate extreme of "zero risk" merely assures mediocrity. That is, the more you invest in tomorrow, the less the stock market likes you today. This over-fascination with the immediate results of any action or business decision has kept the U.S. in sharp contrast to Japanese corporations for many years. Yet, it is interesting to recollect that major development projects like the Manhattan Project and NASA's efforts aimed toward landing on the moon, with their long-term focus, delays, disappointments, and cost overruns, were not abandoned due to their unique context. The overriding consideration in both cases was the earliest completion at (almost) any cost—with the Manhattan Project used to win World War II and the moon landing project symbolizing a prestige effort intended to overtake the Russians, whose own space program took an early lead over the U.S.

Yet, even in the midst of such weighty projects, bureaucratic controls may continue to impede our operations, and sometimes, in their quest for procedural perfection, provide comic relief. For the top-secret Manhattan Project, DuPont committed enormous resources on a cost-only basis (no profit) for several years for the government. However, to meet a technical legal objection, the government ruled that DuPont must accept a nominal fee of $1 at the conclusion of the contract. After V-J Day, DuPont received a check for the full fee ($1!), but their internal auditors took strong exception: The contract had not yet concluded. DuPont, therefore, temporarily had to refund $0.33 to the U.S. Treasury! (Augustine, 1994).

While a whimsical example, the above story illustrates a key difficulty in managing projects within many organizations: the effects of bureaucratic encroachment. Indeed, within some companies the term "staff infection" has been coined specifically to describe the problems that surface when excessive bureaucratic controls are put in place to oversee development projects. An organization's staff structure plays an important and necessary role within the overall business operations. Nevertheless, there are some very real problems that begin to assert themselves when an organization allows the same bureaucratic controlling structure to oversee its project management processes.

One of the central difficulties with allowing a firm's bureaucracy to cohabitate with its project development lies in the disparity of their goals. An organization's bureaucratic core is essentially predicated on ensuring the continuation of operations, the maintenance of the status quo, and the smooth flow of information up and down the hierarchy. Contrast these goals with those of a project manager and his team: the disruption of the status quo through the creative process, the short-term nature of the development itself, and the need to encroach on or violate departmental boundaries through soliciting information and resources horizontally, across departments. Too rigid control, as often typified by bureaucratic structures, has a seriously debilitating effect on an organization's ability to exert flexibility and rapid response to environmental threats and opportunities.

A similar point was made some years ago in a ground-breaking book by two American authors, McCann and Galbraith (1981), who argued forcefully that restrictive control mechanisms as personified by company bureaucracies were the bane of innovative firms. While they have their role within any organization and most particularly within firms which face little change in their markets or fluctuating demand for their products (e.g., public utilities), they are seriously counterproductive for companies that must continually seek to exploit transitory market opportunities.

What, then, is the answer? Readers need to seriously observe the degree to which rules and standardized procedures have become a way of life within their own organizations. Chances are, the more formalized the company's operations and the more rigidly it adheres to standard operating procedures, the less likely it is able to innovate in a timely fashion. Is your bureaucracy stifling your response time and market opportunities? If so, maybe it is time to give serious thought to methods for decoupling project management from normal channels of organizational control.

One method for decoupling that became well known in corporate folklore is the creation of separate "Skunkworks" project organizations. "Skunkworks," a term coined by a team of Lockheed engineers in the 1950s who were working on the development of the U-2 spy airplane, are based on the idea that truly innovative projects cannot operate under a bureaucratic cloud of lines of responsibility and excessive control. Instead, they are set up to operate outside the bounds of the traditional structure, given a specific mandate, and the resources to pursue their goals. Used sparingly and supplied with the necessary personnel and capital, many organizations have found their skunkworks programs to be tremendously successful.

"KISS" YOUR PROJECT

Most of us are by now familiar with the acronym KISS, which stands for KEEP IT SIMPLE, STUPID. We, among others, have often exhorted project managers and project engineers to keep project management (as well as project cost control) as simple as possible, so that everyone concerned can understand and take the simple message to heart. To this end, the acronym KISS can be helpful, if only as a reminder of what is truly important, especially in this age of (almost) indiscriminate use of computers and sophisticated techniques. Certainly, computers are an extremely useful tool, but by themselves, they are no answer to the real problems and issues project managers face on a daily basis. Rather, we have found that just an hour-long walk every day by project managers around the project site can give them all the information they require instantaneously.

Questions are answered on the spot and corrective actions can be taken immediately; no meetings, no minutes, no paperwork, just straight forward action. No amount of progress reporting, computerized or otherwise, can give such a clear picture nor prompt instant action, on the spot directly with the persons concerned.

We are reminded of a newly appointed director of a mega project in India who was very proud to show us in the first week of February 1995, a pile of computer printouts for various aspects of project status as of the end of January. We asked rather innocently what it all meant? His answer was illuminating and, unfortunately, not atypical, "I don't know, I haven't had the time to study it, really." The same response was given in the first week of March, of course, with a totally fresh set of printouts for project status as of the end of February 1995. We venture to suggest that such an impressive printout "heap" serves no purpose except to impress strangers and visitors. What the project director really needs is just a one page status report, giving the status of key subactivities and exceptions. For example:

- Delivery of compressor from XYZ is delayed
- While being shipped by sea from Germany, the reactor sat on the open deck and fell overboard during a storm
- Construction in the battery limits is delayed by twelve weeks.

Given such specific problem areas, the project director can promptly take appropriate action, e.g., getting on the phone with his counterparts in India and Germany. In this way, little time is lost as the main status points are not being buried under a mass of minutia—a classic example of the appropriate use of KISS!

COMMUNICATION: SEEMINGLY SIMPLE BUT FRUSTRATINGLY HARD

Hundreds of books have been written on various aspects of this subject, but we feel that we cannot do better than quote from the well-known management consultant, Peter Drucker. Writing a foreword to a book by Parkinson and Rowe (1977), Drucker asserts that for communication to be successful, we must know surprisingly few points:

- What to say
- When to say it
- To whom to say it
- How to say it

So simple, yet so difficult! Information and the instructions concerning the project are conveyed by the project manager through communication, either written or verbal. Communication, therefore, is the vital link between the project manager and his project team, as well as between him and his client, his subcontractors, and any others involved in the project. Effective communications are essential to the successful execution of any project. Yet, in reality, communications within a project are seldom all that effective, as proved by the results of two surveys on this subject (Warner, 1981). We are all being constantly bombarded with over 2,000 messages everyday; of these we are likely to remember perhaps less than 100. Not very encouraging, but perhaps understandable. The mind has a tremendous capacity to filter out "noise" and unwelcome messages such as those provided on television. Unfortunately, the same filtering mechanism makes it difficult for us to determine important messages from the unimportant, leading to a serious problem with effective communication. The other survey confirms this very low rate of information retention, and goes a step further by looking at the percentage retention of messages conveyed by the normal means of communication generally used, e.g.,

Reading: 10% retention
Hearing: 20% retention
Seeing: 30% retention

The combination of hearing and seeing led to 50 percent retention. If this is combined with some action (doing), the retention percentage is likely to go up considerably, perhaps almost approaching 100 percent. No wonder the trick lies in practice; that is, involving subordinates in the decision-making process as a method of information retention rather than simply having project managers telling them what they believe is relevant. It is this close-to-the-chest approach that is a frequent culprit in information misunderstanding.

In the case of projects that forge international relationships, there is yet another challenge that has to be met. In developing a project, the workforce may well be from a number of different countries, all speaking different languages. The problems associated with this are many and varied. Normal speech patterns use a large number of idioms, colloquialisms, and slang expressions which have no meaning whatsoever to a foreigner unless he is very well versed in our language. For instance, the word "scratch" means "to scrape mildly, as with the fingernails," but we also have idiomatic uses of this word. Recently, a company sent a telex overseas saying that certain items would be made "from scratch." The telex reply said: "Think the word scratch should be sketch!" (Axtell, 1985).

BAR CHARTS AND NETWORKS

The Gantt bar chart, popularly known as just the bar chart, is one of the simplest tools for monitoring projects. An initial version may be drawn from memory on the back of an envelope by the project manager, based on his "feel" for the project. It is simple enough to be carried in the project manager's mind and can be recalled with great precision without any papers and/or documentation. So simple, yet most effective.

As project management became more sophisticated, especially with complex and large projects (the so-called megaprojects), several other techniques and networks were developed to meet specific and more challenging requirements. These include CPM (Critical Path Method) and PERT (Program Evaluation and Review Technique). With rapid introduction of computers in almost every phase of human activity, such techniques thrived—in some cases, more for name sake than real value added to the much simpler bar chart. In due course, it led to a "marriage" of the two, creating a huge number of computer-aided project tracking software packages (Lichtenstein, 1988). These packages claimed to make project cost control simpler and better, particularly when the entire project team had access to the game plan. Despite this claim, Lichtenstein warns that monitoring a project's status with them is frequently not easy, and may make project cost control even more difficult. Computers, of course, do facilitate the operation of project monitoring. A bar-net project schedule combines the best elements of the Gantt bar chart, the CPM logic network, and the milestone schedule into an easily understood means for tracking the status of a project. The advantages include:

- Readily available results of corrective action
- Access to key data that works to keep a project on track

A bar-net schedule is a rearranged preliminary timetable with project milestones marked at the appropriate points. A resource-balanced schedule of the project, as planned, is included so as to present on a single diagram all the pertinent information. The effects of a series of queries, "what if . . ." is easily seen at a glance. Although suitable for both large and small projects, the bar-net chart is particularly effective for projects with under 250 activities. In the case of a larger project, one can deal with segments in a series of bar-net schedules.

So far so good. However, reality often sets in quickly when the use of much of the sophisticated project tracking information is closely examined. In our consulting and personal experience, we find time and again that most effective managers are inclined to opt for the simple old-fashioned bar chart. It cannot be beaten for its simplicity, ready understandability, and effectiveness. Many of our clients, including a number of the major international contractors develop CPM, PERT, or the bar-net charts described above, but privately confess that it is often merely for the purpose of impressing the client by demonstrating that they are aware of and even use the latest and advanced techniques for project cost control. But after this purpose has been served and they have won the contract, in practice they stick to the good old bar chart! Need we say any more?

CHANGES COST DEAR

In the previous chapter, we have impressed the idea of freezing designs at the outset and resisting the temptation of changing them mid-stream. Processes and technology are dynamic, and improvements are being introduced all the time. In fact, it is not uncommon for a plant to be obsolete by the time it is completed. But for sake of timely completion, the design must be frozen, otherwise there is no end to it.

It is a well established fact that any change is quite expensive. And the later a change occurs, the more expensive it becomes. Change is disruptive, no matter what the circumstances are. Unfortunately, corporations are often faced with a classic conundrum in attempting to determine the proper response to change orders. Specifications frozen too quickly can lead to projects that miss the mark, and immediately become obsolete or useless. On the other hand, project stakeholders, such as clients left with total freedom to exact change orders, can create situations in which projects spin on and on, as costs and schedules spiral out of control. It is possible to inhibit change by making up our minds as to what is wanted before we start and resolving to stay with those initial decisions, but it seems that very few owners do that. There is hardly a project built that is not subject to change during the course of construction. Nor should all change be avoided. For instance, a mistake may be made in detailed design that when discovered calls for correction, and initial poor definition of a project will always result in changes. While mistakes will never be completely eliminated, changes can be drastically reduced if a step-by-step approach is adopted. For instance:

- Complete the basic project definition at the outset;
- Do not proceed with detailed engineering design and procurement until the planning has been finalized;
- Delay the start on site until detailed design and procurement for the whole project is within six to nine months of completion.

That last provision is subject to the size of the project, but the principle is sound. It seldom saves time to rush on to the site and start work, although that may impress top management and particularly, outsiders. The staged approach recommended above may (seemingly) prolong the time the project takes to complete, although this is questionable. A logical approach to the project should save time overall, since there will be no lost effort. There is a great temptation to telescope the various phases of a project with a view to saving time, but the saving rarely materializes.

Then there are scope changes, to which the managing contractor must give a great deal of attention. He will call them client changes, or perhaps contract changes (changes made to his contract by his client). There must be a complete system within his organization for their recording, estimating, and evaluation in terms of their effect on the contract. Upon receipt of a request from his client, the contractor must assess:

1) Estimated or actual cost

2) Effect on his engineering schedule

3) Effect on project progress

Thereafter, he registers:

4) Acceptance of items 1-3 above by the client

5) Authorization by his client's project manager

The promoter of the project should recognize from the very beginning that scope changes will arise, and he too should set up a parallel and even more elaborate administrative procedure for their close scrutiny and control. The manner in which this is done will vary from case to case, but a form should be prepared that has to be completed within his own organization before a scope change is authorized and passed forward to the managing contractor for action. That form should include the following basic and essential information:

1) Brief description of the change. This should include whenever possible the equipment involved, together with the related design and construction work.

2) Descriptive data, such as drawings, sketches, flow schemes, and equipment lists.

3) Justification for the change being proposed. For example, easier or cheaper operation or maintenance, safety, technical and marketing requirement, alternatively better efficiency and yield, and better quality product.

4) Estimated yearly saving, if any. This is the financial benefit to be expected, and thereafter the benefit/cost ratio. To be justified, it must be well above 1.0.

5) Required start and finish of the work involved, and its effect on the overall completion schedule of the project.

It is also essential to answer the following key questions:

6) Can the plant operate at full design capacity even if the change is not made?

7) Will the plant be forbidden to operate by any outside authority, e.g., regulatory, if the change is not made?

Once this form has been completed, it should be formally approved by the interested parties within the promoter's management organization, until it finally reaches the project manager for acceptance or rejection. He must have the final voice, giving his reasons, of course, if the scope change is indeed rejected.

In general, the larger the change, the higher the impact on the fate of the project. Any really large scope changes must be avoided except as a last resort, because of their huge impact on the project time and cost schedules. The impact of both time and money is usually more significant than it appears at first sight. This can prove to be very expensive and time-consuming. The scope change costing seldom reflects the full, true cost. It is therefore far better to defer major changes to a later stage, completing the project first. They can then be treated as completely separate small projects, which can then be implemented at optimum cost, without causing undue delay in the completion of the major project.

WE CANNOT DIRECT THE WIND, BUT WE CAN ADJUST THE SAIL!

Too much success can go to the head of an individual or organization. You cannot rest on your laurels for too long, lest it promotes rigidity, complacency, and resistance to change. Likewise, too much success can impede an organization's opportunities for learning. An important text by Hartley (1994), *Management Mistakes and Successes,* paints a convincing portrait of the prob-lems that come from ennui: conservatism, conceit, and complacency. Companies that have been successful in the past lose the incentive to continue questioning assumptions and operating in a competitive manner. Instead, they find it easy to substitute a casual disregard for the environment, assuming that what has always worked in the past will continue to yield dividends. Consider, for example, the recent story of IBM, beset by problems of a bloated bureaucracy and the conceit of past successes. A company that until recently was the envy of the industrialized world saw its stock plunge, its sales slow, and its asset value plummet—in little more than eighteen months!

The counterbalancing effect of a little failure can help induce humility and encourage new approaches (Huber and Glick, 1993). When failure does occur, realistically evaluate its causes, resisting the natural tendency to offer up sacrifices by firing personnel. Instead, it is necessary to understand the root cause of the problems, press on for improvements, and try to encourage breakthrough. Instead of blaming any person for the failure, it is best to blame the process. This can be far more constructive and ensure success in the future.

Unfortunately, we are so obsessed with success, that when someone (as is quite normal) makes a mistake in the corporate world, everyone runs for cover. Success is celebrated, but more importantly, the lessons of mistakes should be heeded. Instead of blaming anybody, try to resolve the situation. The Japanese have a saying for this attitude: Fix the problem, not the blame! This has been the philosophy of Bill Gates and is, perhaps the bedrock of his company's phenomenal success. Microsoft's spreadsheet, Multiplan, introduced in 1984, was found to have a bug that could damage data. Gates had no hesitation in making an instant decision that a corrected version be sent to every one of the product's 200,000 customers, free of charge. The cost to Microsoft: $250,000. There was no need for any discussion and no questions were asked. Management's priorities were crystal clear; the decision gave the distinct impression that the company cares about its mistakes and stays on top of fixing problems. Setbacks are normal, especially when trying new things. The right atmosphere (no fear of retribution) encourages creative ideas—so important for long term success (Gates, 1995).

Drawing lessons from mistakes reduces repetition and further compounding of errors. Multiplan's very first version (1981) was soon running on over 100 different computers. Gates' blunder was to insist that it run on Apple II, a popular but underpowered computer. In making this decision, then, Microsoft had to leave out some potential features, not just for the Apple version, but in all versions of the package. Lotus, a new company at the time, did not make the same mistake with its Lotus 1-2-3 which was designed expressly for the more powerful IBM PC; it ran faster than Multiplan. Gates admitted (1995: p. 31): "I aimed Multiplan too low, knocking it out of the spreadsheets."

More recently, Gates made the opposite mistake in regard to Microsoft's version of Word for the Macintosh, requiring more machine resources than a large number of their customers had. As Gates related,

> *"This got us scrambling for a modified version. Only my job security as CEO (of my company) enabled me to do so; not all CEOs may be so lucky! Many of them are being replaced globally in response to shareholders' pressure. Each case may be unique but this trend is generally for the good. After all, they get paid the most and are expected to deliver the goods . . . though it would be good for the long run, that they should be allowed their share of mistakes, especially if they learn from them." (1995: p. 31)*

Perhaps one of the challenges facing Microsoft is that many of its employees have yet to suffer much failure, if any at all. The illusions spread by hubris can lead to complacency in which future success is taken for granted. To safeguard against this, some organizations have deliberately taken on managers with experience in failing companies so they can use their experience to recognize missteps, cope with such situations as and when they occur, and make mid-course corrections to bring the company back in line with its goals. Microsoft's hiring of Craig Mundie in 1992 is an example in this regard. A decade earlier, Mundie cofounded Alliance Computer Systems, an enterprise which failed when the market for super computers disappeared. Mundie became wiser as a result of his mistakes and is widely regarded as one of Microsoft's best assets. Bill Gates says that he would be willing to hire ten more like him, if only he could find them! Notice the importance placed on lessons to be learned from failures. In the ultimate analysis it is really the failures which force us to be creative, to dig deep, and think hard. Every company should have people who have failed but learned from it. Is there a moral here? Of course: make the most of your mistakes, rather than lament bad fortunes. It is an integral part of life, whether as an individual or as a corporate body.

Ultimately, the Proof Is in the Pudding

We have attempted to highlight a few aspects of the many that we have seen playing a role in cost and time overruns during the course of building major industrial projects, but we feel that the best lesson lies in the case studies themselves. Practical examples are always better than theoretical studies. Unfortunately, every project is different, so it is highly unlikely that we can say with confidence: "I have seen this situation before." While experience is undoubtedly the key to success, that experience cannot tell us what we may have to meet on a new project. What it can do, however, is enable us to handle the situation effectively, knowing what may happen. This recognition allows us to frequently avoid the pitfalls that were encountered last time, assuming we have learned the hard lessons! Experience is indeed the key to success, and our objective has been to broaden the experience of our readers by laying hold of the experiences of others, and learning the lessons that only come "the hard way."

The most important point of all, we believe, is a realization of the profound significance of those first, early estimates of time and cost, especially when they are used for the authorization of funds. It is inevitable that those estimates serve as a standard of reference in all subsequent discussion, however faulty they may be. Unfortunately, it is of no use to say later, "It was only an estimate." Likewise, it is of no use to allege that not enough time had been spent on the development of the project, or that the scope had changed. These are all seen as mere excuses, and are likely to simply be brushed to one side. All that matters is that the original estimates have been far exceeded, and the project is heading toward disaster. Management must then ensure by whatever means open to it that those early estimates are sound, based on a completed basic design.

From then on it is crucial that every change in scope is identified, scrutinized, recognized, authorized, and publicized! For each change, the question should be asked, "Must we have it?" Then, and only then, will management know where they are going with the project. This point is important enough to bear any amount of repetition.

References

Augustine, N.R. 1994. "On Practical Engineering," *The Bridge* (National Association of Engineers), vol. 24 (3), pp. 22–26.

Axtell, R.C. 1985. "Watch your language overseas," *Chemical Engineering*, vol. 92, May, pp. 135–138.

Gates, W. 1995. "The Importance of Mistakes," *Business Today*, May 22-Jun 6, p. 31.

Hartley, R.F. 1994. *Management Mistakes and Successes*, 4th Ed. New York: Wiley.

Huber, G.P. and Glick, W.H., eds. 1993. *Organizational Change and Redesign—Ideas and Insights for Improving Performance.* New York: Oxford University Press.

Kharbanda, O.P., Stallworthy, E.A. and Williams, L.F. 1987. *Project Cost Control in Action, 2nd Edition*, Aldershot, UK: Gower.

Lichtenstein, C.W. 1988. "Bar-Net Schedule Eases Project Control," *Chemical Engineering*, vol. 95, March 28, pp. 53–56.

McCann, J. and Galbraith, J.R. 1981. "Interdepartmental relations," in *Handbook of Organizational Design, Vol. 2*, P.C. Nystrom and W.H. Starbuck (Eds.), Oxford University Press, pp. 60–84.

Montgomery, R.L. 1981. "Listening made easy," Review in *Nations Business*, October 1981, pp. 65.

Parkinson, C.N. and Rowe, N. 1977. *Communicate—Parkinson's Formula for Business Survival.* New York: Prentice-Hall.

Warner, H.W. 1981. "The Reader, Listener, Viewer," *Vital Speeches of the Day*, Vol. 47, pp. 603–606.

Part 2

The Developing World Can Least Afford It

BY "LEAST AFFORD IT," we mean that the developing world is in the worst position to attempt to learn from their failures because of their limited resources, particularly in scarce foreign exchange. Many of the problems encountered in developing countries that are striving to complete projects successfully may be due to their cultural traditions, their ability to secure funding for large investments, and the willingness of governmental officials to operate in a noncorrupt manner.

The projects included in this section are typical of many in the developing world, including the countries of Asia, South America, Africa, and the Far East. It seems that political interference of one sort or another is common in such countries, and this almost inevitably leads to disaster, as money is channelled into other directions and schedules lag far behind projections.

REQUIRED: A BANK OF BLUNDERS

To err is human, but to forget the lessons is certainly not good business. One common thread that we will touch on again and again in this section is how some third world countries have often made mistakes in large projects, particularly when they were the first of their type. Nevertheless, there is often a refreshing willingness on the part of these countries to take an honest look at how the problems developed and learn from those mistakes for future projects. These cases should be viewed as just another resource for foreign project managers or those who have contracted to develop projects in developing countries. Our advice to these individuals is to look to develop a set of lessons from other mistakes in similar settings: In effect, the development of a bank of past blunders that can be used to shape the direction of future efforts. As one writer suggests, there are benefits from consciously collecting past mistakes (Singh, 1994).

The challenges associated with conducting major projects in the developing world are unique. In most situations, the countries do not have the infrastructure to support such projects; they may not have the reserves to pay for them (particularly in a timely fashion), and coupled with cultural misunderstandings and different perspectives, creating a constructive multinational relationship is often an art as much as a skill. At the same time, project expertise is vital to these countries because of their limited capacity for absorbing losses and project failures. Should the Channel Tunnel project fall into insolvency or (heaven forbid) suffer a catastrophic failure, though painful, it would be far easier for British companies to swallow the losses, even to the tune of $15 billion dollars. On the other hand, a catastrophic failure of a large plant or infrastructure project in a developing country can, quite literally, ruin the economy of that country for years to come.

BEWARE, "WHITE ELEPHANT" PROJECTS

Projects are the essential for the economic development of a country. But it is crucial to assess their cost/benefit analysis to ensure that the funds are well utilized. A recent book on the subject from an academician at the Bradford University Development and Project Planning Centre, explains in detail the technique to assess the economic impact of planning in developing countries. To facilitate the use of project analysis as a planning technique, the authors suggest:

- Setting up national economic parameters for use in project analysis
- Simplifying the project analysis methodology
- Helping to ensure consistency across analysis of different projects
- Refining techniques to estimate sets of national economic parameters to help minimize cost of estimation (Curry and Weiss, 1993)

A distinguished banker, John L. Moore, Jr., one time president of the U.S. Export-Import (known as "Eximbank") bank (1977–1981), and later, a member of the board of directors for Bechtel, has discussed his experiences with funding large-scale development projects in the developing world. Based on his experience, he notes (Moore, 1989: p. 305):

> *Project lending and privatization . . . need to be examined in the light of both lending by bilateral official lending agencies, and whether privatization and project finance of infrastructure projects are really of long-term benefit to middle-income developing countries. Personal experience over many years leads the author to caution that these projects must be carefully examined. Without a fair balancing of risk between country and developer, the end result is likely to be white elephant projects on a gigantic scale.*

Moore argues that several variables must be considered prior to committing large funding to overseas projects. First and obviously, can the lender country afford the cost—in particular, should worse come to worst, can the lender country afford the loss? This leads to the second point: Lenders must be willing to accept a reasonable amount of risk in their investments. The long-term payoffs often associated with foreign investment make it nearly impossible to chart out clear liquidation schedules and pay-out rates. Does the lender understand the nature of the inherent risk? Third, foreign investment often comes about through heavy political pressure on the rich lender countries. As such, it is frequently difficult to "call in markers" at a speed that the lender would like. In effect, the problem becomes a political one even more than an economic one. Finally, Moore cautions against the urge to over-lend, particularly without adequate safeguards built into place for monitoring expenditures. He notes from past experience that in their zeal to lend money for third-world development projects, many countries lose control of the money, pouring billions into regions and governments that squander it on worthless or ill-conceived ventures.

In the late 1970s, Eximbank (the precursor of the World Bank) loan officers supported only such projects that really benefited the country's economy, unlike some of the suppliers in Europe, whose main consideration was to sign up export orders. The result has been a slowdown in recent years, in the number of worthless projects that were initiated simply because money was available to start them. Nevertheless, third-world investment will continue to be a tricky place for investment.

DOES AID WORK, REALLY?

Despite severe criticism and general public perception, it seems that the majority of aid does achieve its objective, but with an important proviso. Normally the evaluation of foreign aid packages and their outcome is done just after project completion. But an important study on the subject by an academician at the International Development Centre, University of Oxford, suggests that evaluation should be done at least two years after project completion (Cassen, 1994). It is well-known that any project has a gestation period before the real benefits are apparent, though their full realization may take a much longer period. Cassen notes (p. 236):

> *An enormous amount of project experience has been given and documented . . . but when known mistakes are repeated or success not followed, there is a strong case for effort to improve the learning process.*

In view of the changes happening in many developing countries, it is necessary to make a fresh attempt on how to create a "brave new world" overseas. At the same time, the lessons of the past project failures and the squandering of aid must be learned in order to chart a proper course for future development. A detailed analysis of development in many third-world societies vividly illustrates the difficulties of combining growth and equity objectives while trying to resolve their inherent dilemmas.

References

Robert Cassen, Assocs. 1994. *Does Aid Work?—Report to an Intergovernmental Task Force, 2nd ed.,"* Oxford, UK: Clarendon Press.

Curry, S. and Weiss, J. 1993. *Project Analysis in Developing Countries.* London: Macmillan.

Moore, John L, Jr. 1989. "The Experience of the Bilateral Official Lenders in Funding Large-Scale Development Projects in the Third World," *Technology in Society*, Vol 11, pp. 307–314.

Singer, H.W. and Sumit, R. 1993. *Economic Progress & Prospects in the Third World.* Aldershot, UK: Edwards Elgar.

Singh, J.P. 1994. "Making the Most of Mistakes," *Economic Times,* September 16, Esquire Section, p. IV.

7

Third World Macroprojects

SINCE THE 1970S, there has been enormous investment activity in the developing world undertaken by multinationals with considerable experience with projects in the developed world. But the projects executed by them in the developing countries have not always gone smoothly due to a variety of reasons, including the enormous gulf in social and cultural values and modes of operation. A striking aspect of these failures was the human side of the collaborations. This chapter will address the important question: How can apparently disparate participants work together effectively?

What is a macroproject? They are, at times, also referred to as mega projects. Usually a project costing a billion dollars or more qualifies for this distinction. Because of the tremendous budget and significance of their results, macroprojects can have a huge impact on an organization's, and indeed, a country's, welfare. Hence, there is temptation on the part of political leaders in many countries undertaking macroprojects to "lend assistance" as necessary. Unfortunately, what one party may call "assistance" often seems more like direct interference or overt meddling to project developers. One powerful lesson that consistently emerges from our study of macroprojects is the following: Projects and politics do not mix, or perhaps more correctly, they should not be mixed. But mix they do in real life, and the mixture often leads to disaster. Here, we bring you some real-life examples.

PROJECTS AND POLITICS SHOULD NOT MIX!

Governmental interference in project development comes in all shapes and sizes and often, from quarters that were rarely considered. The examples we bring you here all relate to British aid for dubious projects in return for buying British arms. Both actions end up in disaster, literally and figuratively! Certainly, Britain is not alone in this regard. The U.S. has its own history of offering aid and technical project assistance, with strings attached. These relationships can have detrimental effects for both the developing and developed countries.

Developing countries need a variety of projects, some of them macro, for their quick development, particularly in the areas of infrastructure improvement and by means of raising standards of living. For this process, they need foreign technology and frequently, capital. In times of recession as well as normalcy, developed countries likewise need project orders to keep their economy going. Combine the two and you have a seemingly ideal situation. Unfortunately, in their zeal and ambition, the developed countries often like to tie such deals to additional contracts such as selling arms, giving their "home" industry a further boost. One has only to look around the world and they will discover this pattern repeating time and again. If the project happens to be of dubious character and it is tied up with arms, no wonder we end up with disasters all around. The following paragraphs describe a few examples to illustrate our point (Anderson, 1994).

A British package of 234 million pounds in aid for the construction of the Pergau dam in Malaysia in 1989 was granted by then Prime Minister Margaret Thatcher. This deal was ostensibly a "sweetener" for a larger, $1.3 billion dollar arms order which she personally negotiated with Malaysian prime minister, Mahathir Mohammad. The first installment for this aid was approved despite advice from Sir Tim Lankester, then permanent secretary at the Overseas Development Administration, who argued that the linked-package was "an abuse of the aid program." The World Development Movement (WDM), the small pressure group in Britain, took this issue to the High Court, who ruled toward the end of 1994 that the government did indeed act illegally in granting the aid for the Pergau dam, since this project was not promoting the development or maintaining the economy of a country or territory outside the U.K., or the welfare of its people, as per the 1980 Overseas Development and Cooperation Act.

In the light of the above noted High Court decision, the WDM feels that there is a good case for looking at some similar "shady" project aid deals tied with arms agreements in Indonesia, Thailand, Oman, and Jordan in the last fifteen years since the passage of the 1980 Act.

- An 81 million pounds soft loan to build a power station and a road project in Samarinda, East Kalimantan in anticipation of the Indonesian dictatorship's agreement to immediately buy twenty-four Hawk trainer aircraft plus an additional sixteen aircraft at a later point. These warplanes are intended for use in ground attack roles in illegally occupied East Timor. Apparently, the power station is of an experimental design, and it is not really needed where it is being built. Aid to Indonesia has more than doubled in the last decade in tune with the sale of British arms.

- In Thailand, marked peaks in British aid in 1984, 1989, and 1992 all coincide with massive arms deals. The projects to which the aid was directed included: a much criticized 3.2 million pounds power project won by Balfour Beatty (the main Pergau contractor) and a giant and ecologically dubi-

> ous irrigation and hydroelectric power scheme. Fortuitously, the latter aid was canceled since the corresponding arms deal fell through.

The obvious problem with such linkages is that foreign aid for seemingly necessary capital and infrastructure improvement projects is frequently linked to the true contract, armament sales. As a result, in a number of developing countries, projects of limited utility are being initiated as bait for the sale of weapons. Not only are many of these projects not needed, but they skew the purpose of foreign aid. Mega projects are vital for these countries, but they must be the right mega projects, put into the right use, at the right time, for the right price. As deal sweeteners, we are seeing increasing numbers of such projects being developed not for any tangible purpose in themselves, but as ethical and moral blankets thrown over the more devious underlying contracts.

Arms for mega projects is a trade-off that cannot continue indefinitely. It is creating a tremendous drain on capital in these developing countries as they are lured to spend precious capital reserves on contracts that have no long-term benefit for their people. Meanwhile, many of the projects that do get developed turn out to be unnecessary or ecological nightmares, precisely because they were throw-ins with no well thought out purpose in their own right.

COMMUNICATION: THE CRUX

On projects, as in almost every walk of life, effective communication is the crux. This is particularly so in the developing world, where the infrastructural facilities are rather poor. One of the authors remembers quoting/executing for a project in the 1960s at an iron ore mine in a remote site in India. No telephone or telex, let alone fax service was available. Hence, one of our major concerns was that of open communications with the home office. To gauge the current level of communications, we set up an experiment.

Every day for about a week, we sent a telegram and a post card to our client in care of the project site and requested them to do likewise in the reverse direction. The results were not encouraging, but were, unfortunately, typical of such situations: Telegrams took three to four days in transit and postcards up to ten days. Our first decision was to install a daily courier service in either direction by train, where the journey time was about twenty-four hours.

In another instance, one of the authors was recently discussing the problems of maintaining constructive links to field projects in developing countries with the manager of a billion dollar construction project. The project manager's company was the prime contractor in a mega project in an Asian country, using local workers and subcontractors. He admitted that because of company cutbacks, he was unable to travel overseas more than once a month on direct supervisory trips. Because of a primitive communications setup, the rest of the time he had to use telephones or express mail. As a result, he discovered that every report he received from the site was so optimistic and sanitized that they bore no relation to the actual conditions he would discover when he could finally visit. Clearly, he found that he was the project manager in name only as he had no practical or useful communications and monitoring capacity.

In sharp contrast, some thirty years later we now read of the development of an ultramodern communication system in connection with a major, $2 billion refinery construction project in Venezuela, another developing country, half a world away (Menzies, 1994). This unique, satellite-linked, real-time, customized communication system ties together all the concerned parties for the major refinery project. The owner, Marven S.A., an affiliate of PDVSA (Petroleos de Venezuela S.A.), contracted to build a 320,000-bpd refinery on the Paraguana peninsula in western Venezuela. The work was divided into four contract packages based on technology processes or support functions required for the expansion project. There were four primary contractors:

- M.W. Kellogg Co., of Houston, Texas and its Venezuelan partner, Inelectra. Their responsibilities included engineering construction for all offsites, utility, and infrastructure, and overall program management

- Fluor Daniel and its Venezuelan partner, Tecno Fluor
- Bechtel and its Venezuelan partner, Jantesa
- Foster Wheeler with Otepi

The four contractors with their respective Venezuelan partners concurrently set up eleven project offices for execution of this complex major project. To give a sense of the complexity in maintaining solid communications, consider the site locations:

- Program management team's headquarters in Caracas
- On-site offices in Punta Cardon
- Stateside offices of the U.S. companies
- Caracas office of Venezuelan partners

Each of these offices required interlinking access to the client, and also to each other via the master communication network.

The "One Project, One Team" concept constituted the underlying philosophy, and the communication network was considered as the vital tool to keep this philosophy alive in practice. From its Houston headquarters, Kellogg initially focused on voice and data portions of the network, called VENNET, to ensure telephone, telefax, and electronic mail services. Soon afterwards, these were broadened to include videoconferencing in order to fully utilize the potential of the system. The goals of VENNET right from the conceptual stage were:

- To provide all project offices with multimedia resources
- To offer users immediacy, efficiency, and reliability

VENNET proved to be extremely valuable in speeding up decisions and provided means for asking more players for their respective inputs. Overall, it helped improve team strength, with continual member involvement. It has proved highly efficient and cost-effective, cutting down team members' need to travel. The immediacy of decision-making has considerably facilitated project execution in the shortest possible time. On project completion, the system will be handed over to the Venezuelan owner for use on other strategic projects. On the whole, the system has proved to be of great value and the money well worth spent.

In due course, it may well become a role model for future projects worldwide, especially with the shrinking of the global "village."

BENCHMARKING AND REENGINEERING ENTER PROJECT MANAGEMENT

The above two terms are important concepts in the field of management and as such, are likely to safely survive the "buzzword" status afforded to so many in-vogue business concepts. Although the words originated in the field of manufacturing, it is significant that they are being introduced in the field of engineering and construction projects. So far these functions have received little attention in the project management field and perhaps none at all in case of projects in the developing countries where they are needed all the more. However, the words seem to mean different things to different people, so we should start with their definitions: Reengineering is defined as:

> *"the fundamental rethinking and radical redesign of business processes to achieve dramatic improvements in critical, contemporary measures of performance such as cost, quality, service, and speed."*
>
> (Hammer and Champy, 1994)

For the Engineering/Construction (E/C) industry, reengineering suggests the following actions:

- Reduce debt, broaden capital base, and develop a strong balance sheet
- Improve communications, maximize computer usage, and globalize operations
- Broaden the spectrum of services offered and the customer base and reduce risks
- Use a project-team approach; maintain a flat organization
- Closely monitor all costs and be flexible in all functions (Munsteri, 1994)

In Munsteri's research, he found that out of the ten major E/Cs studied, only three appear to be implementing the above programs. Four others have the same objectives, but their implementation lacks cohesiveness. Reengineering is, in a nutshell, the key to survival. For viability, E/C companies must cater to all industries. The type of companies likely to survive will be (broadly):

- Large, world-class, publicly held, or private companies
- Divisions/subsidiaries of larger corporations that are financially secure
- Medium-sized companies owned by larger organizations
- Small high-technology or regional firms, with costs lower than larger firms

A common denominator of the above organizations is, admittedly, a secure financial base that allows them to make the necessary initial moves toward a reengineering philosophy. Quick fixes and partial steps do not work. As with any systematic change in operations, a half-hearted attempt or one made with many conditional attachments is usually worse than not having attempted the change. Reengineering is a process and a mindset that pays huge dividends, but only to those firms that are committed to the new course. Above all, for continuous survival, the companies must be in a constant state of reengineering; that is, reengineering is a continuous process, much like the total quality philosophy. It is not an end state that can be recognized and positively reached. Rather, reengineering requires organizations to continually test their assumptions and look for increasingly innovative ways to "do it right," and achieve competitive advantage through a higher commitment than their competitors.

Unlike reengineering, the benchmarking process consists of a formal method for companies measuring their operations' or products' service with the best performed in the area. In essence, benchmarking, which is becoming hugely popular in the 1990s (Altany, 1991), is a form of legal copying of the industry standard (Main, 1993). At least, the philosophy gives companies the ability to stake their products against the best in order to determine where they come up short and to make necessary corrections and modifications (Bogan and English, 1994).

Benchmarking for project management was discussed at length by the so-called "buzz groups" from among the over 400 participants of the AIChE (American Institute of Chemical Engineers) 26th Annual Engineering and Construction Conference in late 1994. Comprised mainly of individuals in the E/Cs, their findings on benchmarking in projects are summed up as follows (Vervalin, 1994):

- Critical factors for success are cost consciousness, safety, schedule, and client satisfaction. Any new product or service must comprise, at a minimum, these properties.
- Characteristics are risk, milestones, safety incentives, and management. The basis for an effective benchmarking program is to attend most closely to those factors which can be directly influenced by the company's actions.
- Opportunities are time/cost reduction, operation/reliability improvement. Benchmarking offers a number of useful and productive results.

- Is consensus possible? Yes, if there is a clear program of top-down selling, system-wide commitment, and active participation at all levels.

PROJECT COST CONTROL: AN EXERCISE IN FUTILITY?

One of the authors carried out an intensive consultancy assignment back in the early eighties for the United Nations Industrial Development Organization. This assignment arose from UNIDO's concern about the rapid escalation in the cost of fertilizer projects, particularly in the developing countries. Essentially, there had long been a well-founded suspicion that contractors in developed countries were using the construction of important fertilizer plant projects as a method for enriching themselves well out of proportion to the original terms of the contract. The suspected culprits were a combination of inexperienced third-world governmental bidding agencies and unscrupulous contractors eager to exploit these opportunities.

The exact terms of reference (1981) as well as the final reports (Kharbanda, 1983) were couched in far too diplomatic a language to have any real impact or solution to this serious problem. We bring you here a first hand account of this exercise, including some blunt truths, to illustrate the depth of the problem and hopefully, to point to possible solutions. The basic question was, in effect: Are the international contractors taking the third world for a "ride?"

Nearly two years of intensive work, visits, and inputs from scores of contractors, owners, and consultants worldwide led us to the following honest answer and the possible solution: Yes, due largely to ignorance and lack of homework by the owners.

We have space here to provide our readers with barely the highlights of the two year exercise done at Bombay, Vienna, major capitals, and other cities worldwide, and various project sites in the third world. However, the report itself offered some damning evidence (Kharbanda, 1982):

> *"International contractors are out to make profit, the higher the better. This is their main objective and there is no secret about it . . . An incomplete and a vague inquiry, (as is usually the case) is almost like a signed blank check . . . The winning lowest bid is likely to have many 'ifs' and 'buts' . . . their cost can be enormous, even of the same order as the main price!"*

Is it any wonder then, that the third world owner is frequently cheated? The governments, unskilled in the methods for soliciting precise bids, leave the door wide open to contractors who take advantage of nebulous wording and unclear specifications to produce enormous cost overruns. And the moral? The project owners pay for their ignorance, and the more unscrupulous contractors take full advantage of it. Blame lies with both. Usually a good, reputable contractor may well refuse to quote against a vague inquiry because he values his reputation far more than the short term gains from a project which is almost foredoomed to go sour. After all, nearly half, if not more of the ongoing workload of a reputable contractor is comprised of repeat orders from satisfied customers. In-depth discussions with international major contractors active in the fertilizer field, revealed some useful pointers:

- Thorough homework by the owner is a must: an extra dollar spent on research may yield a million dollar benefit!
- Small and simple plants at user points are far more efficient and cost effective than large, centralized plants
- Use a turnkey approach with total responsibility resting on the contractor's shoulders until the plant is turned over
- Concentrate on the vital few aspects of the specifications instead of on the trivial many
- Time is money. Quick decisions/action can save enormous amounts of both
- Use a detailed invitation to tender bids, use global tendering, and maintain untied credits

PROJECT ANALYSIS AND EVALUATION

We have consistently emphasized the absolute and urgent need to do thorough homework before embarking on a project of any size. The initial expense can save enormous time and money downstream. Let us illustrate this point with a couple of examples.

In China we are told that the procedures for allocating materials and funds in major projects are well established. In the late 1980s, systematic analysis methodology for selecting projects was introduced. Some of the recent reforms in China in this field include: decentralization of decision authority, and financial and price control systems. A manual has been published with detailed financial and economic calculations for illustration purposes. The main purpose of the project manual is to the use consistent forms of project analysis and broaden its institutional basis.

It was found that the main obstacle in extending and applying this process in practice was the dissemination of such information and techniques to the appropriate levels of the concerned state enterprises and their respective ministries in the government (Curvy, 1990). There seems to be an urgent need for broadening the sectoral coverage of project analysis and detailed project calculations from the industrial sector to the agriculture-related and transport sector.

Recent experience in the matter of investment in China shows that emphasis has been more on expansion of existing establishments rather than starting new units at "greenfield" sites without much infrastructure. A strong argument can be made that this represents a very pragmatic move, since expansion is much quicker and also shows more impact on the economic bottom line than developing fresh units starting from scratch. Investment proposals are handled by three main departments in the government: construction projects, updating projects, and foreign investment. These are then referred to the Fixed Assets Department for evaluation and final inclusion in the overall investment plan.

The existing emphasis on industrial investment has revealed a lack of capacity in port facilities and sea transport. Rail, road, and water transport are each located in different ministries with no consistency between methods of analysis. The development of investment analysis in transport needs to take into account the full benefits of transport investments as well as the opportunity costs. It may be that the application of project analysis develops along with the development of markets in productive economic spheres, as there is no real contradiction. However, project analysis cannot develop to its highest potential without reforms in the spheres of production and circulation. Its basic principle is to make stated actions, particularly with respect to investments, more effective.

Our other example(s) relate to the formerly socialist and other developing countries based on the proceedings of a conference on project appraisal and planning (Weiss, 1990). Discussions at the conference included cost-benefit analysis techniques with particular reference to their past and potential applications in socialist economies. These papers discuss the underlying causes for failure of some of the development projects in developing countries, although the emphasis seems to be confined to assessing only the financial/economic viability of projects rather than the project in its entirety. The conference discussants agreed that overall, there is a strong need for developing a systematic approach towards project planning. Further, there seems to be an urgent need to liase between the macro and micro levels of planning, and to expand substantially the project planning capacity of developing countries. The practical implication of current practice is that there is little coordination between the project planning activities and their relationship with overall economic development. Too many decisions are being made at project sites without any coordination with central governmental policy. Addressing this condition requires an administrative agency with appropriate links to the decision-makers at macro-level and the project planners in executive agencies.

FAILURE CAN BE AVERTED, EVEN IN DEVELOPING COUNTRIES

While this chapter is devoted to a variety of case studies of failures worldwide, let us conclude in a positive context. Failure can be averted even in the developing countries. To illustrate, consider the case of a major project, a wellhead offshore platform for India's ONGC (Oil & Natural Gas Corporation Limited). This case is indicative of how cost and time overruns, which are otherwise so common in large projects, can indeed be avoided simply and quite effectively.

The project comprised four wellhead offshore platforms, each about 600 tons with a total value of $125 million. Awarded to one of the most respected project contractors in India, Larsen & Toubro, Ltd., the work started in mid 1992, with scheduled completion for April 1994. Halfway through the project (mid 1993), it had already slipped badly and in normal course, the project would have been delayed by perhaps a year, with serious consequences to ONGC, L&T, and India. Why to India? Nearly half of India's crude oil requirement is imported and the delay in this project would result in increased imports amidst a serious foreign exchange situation. Obviously, the stakes were very high as costs were threatening to spiral out of control. The situation called for immediate and radical action. The primary causes for slippage included delayed decisions by ONGC and the poor organization setup by L&T.

Into this precarious situation stepped two key players in the project's future: Mr. R. L. Samy, deputy general manager of ONGC, and Mr. K. Venkataraman, L&T's general manager. The actions they were to take over the next months led to a miraculous transformation. Their solutions were simple but powerful: Look beyond the theoretical approach and the contract; evolve a down-to-earth and common sense approach to handling problems at hand, and spend time on-site solving those problems. This philosophy was translated to suggest the following main points (Kelkar, 1994);

- Solve the problems at the site itself; don't pass the buck on to headquarters
- The owner and contractor are expected to approach the project as partners, not as adversaries
- Regular meetings, minimum paperwork, on-the-spot decisions

- Necessary leadership at the top for total commitment to mutual benefit of contractor and owner
- Centralize responsibility with two or three persons; make them accountable
- Constant follow-up with suppliers, especially overseas

What was the result of this shift in philosophy? Within six months, there was a dramatic sea-change. The job not only caught up with the original tight schedule, but was actually running a few days ahead. Morale improved, giving a further boost to the project's productivity. The project was commissioned in March 1994, a few weeks ahead of schedule. Both ONGC and L&T not only looked after their respective contractual obligations but also helped each other in times of trouble, realizing that offering such assistance was for their mutual benefit. Problems and schedule delays that L&T faced due to bureaucracy were sorted out by ONGC. To expedite crucial project materials shipments, either Indian or imported, L&T stationed a senior person at the supplier's office or manufacturing shop. Top level weekly and even daily meetings at the site helped resolve problems, and wherever applicable, cost adjustments were agreed to on the spot, equitably and fairly.

Though both L&T, and particularly ONGC, had their respective layers of bureaucracy and rigid systems internally, their project teams were kept nearly flat with little or no hierarchy to create any sort of bottlenecks. Bureaucratic control was minimized and budgets were held outside of normal oversight channels whenever possible. Above all, development decisions had to be quick—quite often almost instantaneous. Both parties had come to realize that time was of the essence and further, that time truly was money. Both Samy and Venkataraman learned the hard way that to do things quickly and efficiently, they must work together as a team and communicate openly and honestly as partners: ONGC + L&T rather than ONGC vs L&T. In this approach alone lay their salvation of this crucial job. As a result, everybody won, including India. Venkataraman has obviously learned a valuable lesson, which he expressed candidly: What is essential for the project to come off well is a demanding client, and ONGC was just that!

That result is interesting and highly instructive. It is common for contractors to think of themselves as at least as smart as the owner for whom they are working. In fact, this bias seems to be the essence of the UNIDO problems we have briefly dealt with earlier. But that is a very short-sighted approach. It may work for one job, but that may be the last one! And that is not what really good contractors thrive on. Obviously, Venkataraman had this point in mind.

Not only did L&T learn the appropriate lessons from this experience, but they retained that information, as future events would demonstrate. Even before the successful completion of the first project, ONGC was satisfied with L&T's performance and attitude, and awarded the contractors a repeat job in January 1994 on a crashed schedule. The $80 million job comprised three wellhead platforms, each 600 tons and involving about two and a half miles of deck piping and the erection of seventeen major items of equipment. The scope of work included detailed engineering, global sourcing, and round-the-clock fabrication at L&T's heavy fabrication shop at Hazira in Gujarat. The decks were ready for load-out in eleven months, some four to six weeks ahead of schedule (L&T Pragati, 1995), despite the heavy rains, floods, and the so-called "plague" epidemic that affected the shop area. This time compares very favorably with international standards and places L&T in a unique position to meet the challenges of turnkey execution in South Asia.

The bottom line is important: The third world presents a unique set of challenges in contracting and developing mega projects that almost presupposes disastrous overruns in budget and schedule. Poor communications, inadequate oversight, occasionally unscrupulous contractors, and excessive bureaucratic controls all combine to offer project organizations a daunting task. This chapter has focused on this challenge and attempted to share some successes as well as failures, to indicate that while difficult, these problems can be surmounted with canny management skills and a clear understanding of what is to be expected in competing in the third world arena. Mega projects, under the best of circumstances, are difficult to effectively manage. Given the external constraints imposed in third world countries, they are that much more difficult. They are, however, not impossible. Foresight and creative solutions are often the key to success here. Where there are no rules, the timid often flounder, but the strong flourish.

References

Altany, D. 1991. "The hottest new buzzword: Benchmarking," *World Executive Digest*, May, 51+, 3 pages.

Anderson, P. 1994. "Pergau: Tip of arms-for-aid iceberg," *New Statesman and Society*, vol. 7, Nov 18, pp. 7–8.

Bogan, C.E. and M.J. English. 1994. *Benchmarking for Best Practices: Winning Through Innovative Adaptation.* NY: McGraw-Hill.

Curvy, S. 1990. "Evaluation of project analysis in China," *Project Appraisal*, March, pp. 2–9.

Hammer, M. and Champy, J. 1994. *Re-Engineering the Corporation.* New York: HarperCollins, 1994.

Kharbanda, O.P. 1982. *Capital Cost Minimization of Fertilizer Proj-ects in Developing Countries*, Interim Report No 2, May 14, Vienna: UNIDO.

Kharbanda, O.P. 1983. *Capital Cost Control of Fertilizer Plants in Developing Countries*, 2 volumes, Vienna: UNIDO.

Kelkar, V.Y. 1994. "Principles of Project Management," *Business Standard*, October 25, the Strategist section, p. 1.

L&T Pragati. 1995. Employee Communication Journal of Hazira Works of Larsen & Toubro Ltd., Jan–Mar, p. 1.

Main, J. 1993. "How to copy ideas legally," *Span,* vol. 49, August, pp. 8–11.

Menzies, J. A. 1994. "How VENNET Improved a Project', *Hydrocarbon Processing,* vol. 73, December, pp. 89–90.

Munisteri, J. G. 1994. "E/Cs: Reengineer for Survival," *Hydrocarbon Processing,* vol. 73, October, pp. 123, 7p.

Vervalin, C.H. 1994. "Views from the '94 ECC Conference," *Hydrocarbon Processing,* vol. 73, December, pp. 86–87.

Weiss, J. 1990. "Project Appraisal in Theory and Practice," *Project Appraisal,* pp. 62–63, Summary of Conference held at the Institute of Planning, Academy of Economics, Poznan, Dec. 14–15, 1989.

8

The Dulhasti Power Project*

INTRODUCTION

This chapter relates the story of the Dulhasti Power Project, a failed project in India. Originally conceived to tap into the vast potential of hydroelectric power, it was a massive undertaking from the beginning. The project was sited in the remote foothills of the Himalayas, leading to a construction schedule disrupted by recurring geological surprises. The project team was burdened with political problems and power struggles between state and central government over control and regulation of the facility. Further, because foreign firms were originally selected to manage the development, there were problems of international cooperation to be sorted out.

---

* Portions of this chapter were adapted from O.P. Kharbanda and E.A. Stallworthy, *Lessons From Project Disasters*, Manchester, UK: MCB University Press, 1992.

As initially conceived, the project's cost was estimated at 1.6 billion rupees (about $55 million). By the time the contract was let, the cost estimates had risen to 4.5 billion rupees and the final cost was over 8 billion (or about $275 million). Clearly, this was a case of a project budget run amok.

With total assets of 70 billion rupees (about $2.3 billion), the National Hydro Power Corporation (NHPC) has a total hydroelectric power capacity of 1638 megawatts (MW). In August 1994, arrangements were reportedly finalized with the French consortium of contractors to resume work. Completion of the project was now expected in 1994, but a revised estimate is yet to be prepared, pending receipt of bids for the remainder of the civil work. Among the many problems associated with this project from the beginning was the absence of a strong project leader, particularly to represent NHPC's concerns. As a result, the project was often held at the mercy of the contractors, who continually dragged out the schedule and cost estimates with little incentive to operate squarely. As this case will demonstrate, the absence of a strong leader can often make all the difference.

HYDROELECTRIC POWER IS VITAL

The importance of hydroelectric (hydel) power for a country like India should be obvious from the fact that fossil fuels, coal, and oil are expensive and polluting, and contribute to the greenhouse effect. Hydroelectric power, on the other hand, offers developing countries a cheap alternative that is virtually limitless. This is particularly true in India, with its broad expanse of waterways and large, rural population. Consequently, Indian governments are justified in their belief that it was imperative to tap India's large hydel potential (Nand, 1995).

Although initiated for the best of purposes, the Dulhasti power project has suffered from huge budget overruns and a seriously lagging schedule. Experiences with the project exemplify the importance of maintaining continuity in terms of design and contracting, although there were many other factors that also caused delays. The goals of the project were simple: Dulhasti was designed as a 390MW power plant to be built in the Doda region in the state of Jammu and Kashmir, a rugged, mountainous region in the Himalayas, several hundred kilometers from larger cities. The project sought to build a dam, erect a hydroelectric generating station, and string hundreds of miles of transmission lines starting near the headwaters of a system of rivers flowing onto the plains south of the mountain region. Unfortunately, the project experienced a variety of problems from the outset. The construction schedule was disrupted by recurring geological surprises and also suffered from political problems, such as a power sharing row between the central and state governments.

A Troubled Past

The contract for the power generation project was first awarded to a Cegelec-led French consortium, who almost immediately asked for an upward price revision despite being given a letter of intent in December 1986, within the time limit they had specified. The Indian government refused, suspecting that the French consortium had known all along that their initial bid was too low and were hoping to simply "buy" the project prior to renegotiating. The government's refusal to revise their price resulted on a second bidding process. Because of wider competition from other European countries now in the field, the second French offer was accepted, even though the bid was lower than their earlier one. Although this process saved the Indian government money, it was not a good start to their partnership with the French consortium.

The second bidding tender concerned the erection of the transmission lines from the power station to populated areas. The winner was a Soviet turnkey bid for the construction of the transmission lines. Unfortunately, their terms included a price of 1.6 billion rupees—nearly five times the initial 1987 estimate. This part of the project was to be funded through the Soviet credit agreement of 1985, because of foreign exchange restraints. Later, the Soviet contractors insisted that the project be financed by commercial credit, for reasons that are not clear. During protracted negotiations over some two years, the Soviet bid was progressively reduced to a final level of 4.5 billion rupees (about $150 million), still nearly three times the initial departmental estimate of 1.6 billion rupees. Bear in mind, however, that because of the excessive nature of the bids in the first contracting round, the project's start had been pushed off over three years. The irony is that by the time the project was initiated, many of the cost estimates that Indian officials had originally generated were most likely no longer accurate, resulting in rapidly inflating budgets, similar to the contractor's earlier bids!

The latest estimate for the power station was now said to be about 2.5 billion rupees, due to a difficult terrain and nightmare logistics. The site of the construction project was impossible to get to via any form other than air. All raw materials, including cement, wood, stone, and steel had to be hauled by helicopter for miles over snowbound areas. Further, the Soviet contribution to the cost of the project had dropped dramatically from the original partnership arrangement. In fact, it had fallen to approximately 10 percent of the projected cost, forcing the Indian government to finance the balance from internal resources. The project has been subject to major delays, both legal (as contracts had to continually be renegotiated or arbitrated) and technical (as start-up delays combined with logistical problems to slow construction) and these delays only brought continuing increases in costs. Meanwhile, given the nightmare that dealing with foreign contractors has become, the concerned Indian company, the National Hydro Power Corporation (NHPC) is wondering whether they can go it alone. If so, they would have to finance the project from internal resources, and having gained much experience in this type of work, it would not be impossible. The know-how for the erection of the transmission lines in the difficult snowbound areas could be obtained from the appropriate companies in Norway or Switzerland, who have the necessary expertise (*Economic Times*, 1991).

These various controversies seem to have resulted in the most serious delays. A report issued in October 1991 stated that while the project was scheduled to be commissioned in less than thirty-five months from its original date of 1985, work has yet to be started on a major section of the transmission line, which is to be executed on a turnkey basis by the former Soviet Union, and is currently scheduled by them for completion in 1997 (*Financial Express,* 1991). This would be an unprecedented situation in power-starved northern India. If a hydroelectric project would come into operation without a transmission line to transmit the power being generated by it, it would inevitably lead to major losses by NHPC, the operating company. And yet, the project is still delayed by disputes over the costs and contractual penalties, and who should pay.

According to official sources, while the Russians are willing to sign the formal agreement undertaking the execution of the transmission network for which they are responsible, the Ministry of Power has not yet authorized NHPC to finalize matters. Clearly, in the wake of their nonperformance to date and penchant for reneging on credit agreements in the past, the Indian government is increasingly reluctant to enter into yet another contractual agreement.

It seems that the Ministry is dissatisfied with the revised estimate submitted by the Technoprome Export Corporation (TEC) of Russia, as it was known at the time of these negotiations. On the other hand, the Russians are firm that cost escalation has raised the cost of the network to 7.2 billion rupees ($240 million) from the previously sanctioned 1.62 billion rupees. The Russian contribution to the costs is said to be only $30 million, and the Indian government is seeking to raise their contribution. It was also argued that while the Russian estimates may be termed too high, there is a lack of local resources to make up for the short-fall. This causes the project to fall under the list of projects that require Rus-sian assistance under the Indo-Soviet protocol in 1986. The result has been a minefield of truly complex financial negotiations that have proved disastrous to the project's progress. What is more, the political upheaval in Russia and its former client states will certainly not help alleviate the situation.

How the Project Has Fared

This total project was originally estimated at 4 billion rupees (or about $130 million). It has become increasingly likely that the Dulhasti project may now end up (if it is ever completed) at over $600 million. By 1992, nearly 8 billion rupees had been spent, and two years later (1994), the French seemed to be bowing out of their contractual commitment, ostensibly for security reasons. Meanwhile, some 5,000 workers on the site have been idle for nearly two years. The French consortium Cegelec Alsthom has served notice on NHPC to make alternative arrangements for completing the project.

An Indian delegation visited Paris at the end of 1993 to try to get the French to resume work against their demand that the Indian government revise the contract upwards by 4 billion rupees and insistence that the French experts should be allowed to operate by remote control rather than work at the building site. The French claimed this demand was due to security reasons. However, it seems that the real reason for French contract intransigence was that they were finding it difficult to bore tunnels through the mountainous region due to serious geological problems that restricted the engineering. The original path had to be diverted. Consequently, there are very real suspicions that security was only an excuse for the French to wriggle out of a contract they have come to view increasingly as unprofitable. The work on Dulhasti has been at a standstill since August 1992 after the abduction by local guerrillas of a French project engineer, Antonio Silva. Meanwhile, the materials and machinery have been allowed to decay in the open atmosphere (*Financial Express,* 1994).

The Dulhasti project offers some important management lessons, primarily in the area of contracting and logistics planning. Unfortunately, due to national political considerations, the NHPC was practically forced to accept a tender bid from a Soviet company to develop the transmission lines for the project. This bid was in the form of a credit agreement between the two countries. However, once well into the project, the Soviet contribution rapidly dwindled, leaving the Indian government with an escalating cost that they could not lay off on their erstwhile project partner. Meanwhile, the price tag for the transmission lines component of the project continued to climb, putting the Indian government and the NHPC in the position of having to renegotiate contracts and consume precious time with legal maneuvering.

The contractual situation was no better where the French consortium was concerned. Their initial bidding posture was at best inauspicious and at worst, fraudulent. Even on the heels of the second bid, which the French consortium won at a much lower price, there were very real suspicions that the contract price would be inflated. At the very least, the problems with contracting stole over three years from the project development time line, increasing costs and wasting valuable equipment and opportunities in the process. Further, due to other technical problems, revisiting and renegotiating contracts has become the preferred method for official foot-dragging as the French desperately seek to be released from their contractual obligations.

A second important lesson has to do with the logistical planning for the project. Clearly, this project was a technical challenge from the outset, given the rugged, remote location for the hydel plant. Did that site add to project delays? Undoubtedly. However, it appears that some of the technical challenges were minimized in the early bidding and development processes. Perhaps a more accurate assessment of project risk analysis, combining an understanding of the logistical challenges, would have led to a more accurate time line and reasonable budget. In its current form, Dulhasti unfortunately became the project that was like a chronically ill patient, one who would neither die nor get well. As a result, to carry the analogy one step further, more and more money was spent in trying to revive an unresponsive patient while the corps of doctors bickered among themselves over the most appropriate treatments.

IS REALITY INDEPENDENT OF TIME?

A major component of any international joint venture lies in a clear understanding of the similarities and differences among cultural groups. India, Pakistan, and perhaps some of the neighboring countries may have a very different concept of time than that prevailing in the Western world. The idea seems to have a spiritual background, associated with the belief that reality or truth is eternal and independent of time. This concept has in the event proved costly in terms of wasted resources and slow progress. It is no wonder, therefore, that delays in projects are the norm rather than the exception, despite the fact that the initial completion schedules are already far longer than is generally acceptable in the developed world (Saha, 1991).

India and Pakistan are typical examples of developing countries. It is interesting to compare India's Hindu culture with Islamic Pakistan. The Indian classical tradition considers time as a mere illusion, whereas Islam treats time as real. Despite this basic difference, the economic performance of Pakistan and India does not seem to differ very much. This fact tends to confound the theory of Saha (1991), who links the Hindu spiritual disposition with India's poor economic performance. Perhaps the problems that exist in India are due to other factors entirely. Kola, for instance, advances a series of very different reasons (Kola, 1991):

- the colonial struggle;
- country in the early stages of capital accumulation;
- the extreme pressure of population growth;
- over-protection of local industry;
- an inefficient and corrupt bureaucratic structure.

However, these factors are equally applicable to Pakistan. Perhaps it would be more appropriate to compare India's performance with that of countries such as Egypt and Turkey. Here there are very obvious differences, despite the fact that these countries have an established religious culture as well. The fact that the Muslims pray five times a day at specific hours does not seem to inhibit their ability to pursue effective economic policies. So perhaps religion, as such, has really nothing to do with it. The fact remains that in India, project overruns are the rule rather than the exception. A detailed analysis in 1987 revealed that about 60 percent of the ongoing mega and major projects, individually costing 1 billion rupees and above, and 70 percent of nearly 200 medium size projects (Rs. 0.2 to 1.0 billion) were running behind schedule anywhere from five months to thirteen years. On the whole, cost overruns erode some 20 percent of the total financial outlay on such projects (*Business World*, 1987).

INFRASTRUCTURE PROJECTS

The funds required for India's current infrastructure projects are mind-boggling. For example, for the vital power sector alone to meet the current targets requires at least 30 billion rupees ($1 billion) every year for the next fifteen years. With barely four proposals cleared, the apex financial institution, IDBI, is already feeling the resource crunch. As of early 1995, there were some seventy-five proposals for about 32,000 MW pending approval by the government. It is feared that sometime during the next five years, funding for the massive infrastructure projects could run dry, thus jeopardizing the industrial production. The main problem lies in the funding of these projects from the domestic capital market.

One suggested solution for funding future infrastructure projects is the development of a bond market. But it is doubtful whether this avenue could yield more than $650 million each year for one sector alone (power). In addition, there are huge demands for other essential sectors like transportation and telephone. In addition to the domestic finance crunch, there are absolute limits on foreign funding as well. Large-scale foreign financing of projects would in due course create a serious balance of payment problems, akin to that experienced by Latin America in the 1980s. The solution may lie in developing marketable bond instruments from India, which can tap the international retail markets instead of limiting current offerings to a few banks (Dutta, 1995).

This approach will require a sea-change in the attitude of the Indian promoters. For example, there is a major problem arising out of the recent agreements, especially in the power sector, where the promoters are different from the operators and the managers. The promoter is guaranteed a minimum return, and any slippage in setting up the plant is to be recovered from the equipment supplier. Likewise, any shortfall in performance is to be recovered from the operator. In effect, the project promoter hardly runs any risk. In the ultimate analysis, the promoter has to prove his credibility to enable raising the debt funds from the retail financial markets.

Overruns Are the Norm

Unfortunately, the position seems to be getting worse rather than better. Not only were the headlines in the Indian press regarding project delays and cost overruns in 1991 very similar to those that were appearing in 1987, but the situation seems to be getting ever more serious (Balashankar, 1991). Balashankar cites 204 delayed projects, with the total original cost first estimated at 342 billion rupees (about $ 11 billion), whereas the revised estimated cost is now over $20 billion and likely to rise much more. Typical examples of such delayed projects are:

	First Estimate Cost in billions (Rs.)	*Completion Date*	*Revised Estimate Cost in billions (Rs.)*	*Completion Date*
Kahlgaon power plant	8.8	July, 1992	15.5	Jan., 1993
Koel Karo Hydro	4.4	Dec., 1988	12.5	March, 1994
Bokaro B-11 TPP	1.9	Oct., 1985	3.6	Ongoing
Haldia Fertilizer plant	0.9	Oct., 1976	6.5	1992
Optical Fiber project	0.3	Oct., 1988	0.5	Jan., 1993
Calcutta Metro	1.4	Dec., 1978	15.2	Jan., 1995

Alleppy-Kayamkulam	1.1	March, 1991	3.6	Not known
Visahapatham steel plant	6.8	Not known	8.1	Not known

The examples we have quoted from the many delayed projects are diverse, both in type of project and size. The Calcutta Metro (one of our case studies) is perhaps unusual, but only because it is being built in India. The cost overruns of oil projects tell their own horror stories with delays in completion of projects ranging from two to three years. Here are a few examples (*Chemical Industry News*, 1995):

	Original cost (billions Rs.)	*Revised cost (billions Rs.)*
Kandla-Bhatinda oil pipeline	9.18	23.92
Neelam Oilfield project	20.22	35.42
Digboi Refinery modernization	1.44	3.64
Catalytic Reformer project at Digboi	.34	1.12
Catalytic Reformer project at Barauni	.78	2.48

Citing these cases, the Standing Committee on Petroleum has called for a thorough revamping of the project planning and implementation system in the public sector oil industry, as their cost rose by over 3.5 billion rupees above the projected figures. Specifically, the concerned Ministry (Petroleum) has been asked to review the project planning and implementation system in order to remove the inherent shortcomings. As a followup action, the Ministry has instructed its nominee directors on the boards of the various public sector companies to keep a close watch on the implementation of ongoing projects.

The Committee has strongly advocated that realistic allocations should be made in the first place to facilitate project execution on time. For example, the Planning Commission approved just half, i.e., 26.5 billion rupees of the 53.2 billion rupees for the Eighth Plan in the Petroleum Ministry. For ongoing projects, full allocation is essential for their timely completion. As with many developing countries, project management in India will continue to require that officials grapple with the effects of runaway budgets. Given these countries' overall inability to absorb significant cost overruns, the sooner that professional and pragmatic project management, particularly of large, capital projects, becomes the norm, the more easily will these countries develop and continually upgrade their national infrastructures.

References

Business Express. 1991. "Rs 30,000 Cr. down the drain," Sept, 10, p. 1.

Business World. 1987. "India's shocking project disasters," March 2–15, p. 42.

Chemical Industry News. 1995. "Concern over cost overruns of oil projects," vol. 40, May, pp. 654–655, original Source: Hindu, Madras.

Dutta, S. 1995. "Is the money running out?," *Business India*, January 16, p. 33.

Economic Times. 1991. "Cost estimates controversy delays Dulhasti project," Feb. 6, p. 12.

Financial Express. 1994. "Dulhasti Project Dogged by Uncertainty," January 18, p. 5.

Financial Express. 1991. "Dulhasti project in dire straits," Oct. 30, p. 7.

Kola, E. 1991. "Traditional Indian concepts of time and its economic consequences, where is the hard evidence?" *Project Appraisal,* vol. 6 (2), pp. 109–113.

Nand, S. 1995. "The Future Lies in Hydel Power," *Business & Political Observer,* March 13, p. 11.

Saha, A. 1991. "Traditional Indian concept of time and its economic consequences," *Project Appraisal,* vol. 6 (2), pp. 113–120.

9

Calcutta Metro: Is the Magic Wearing Off?

THIS CHAPTER TRACKS the history of a huge project to establish comprehensive public transportation in Calcutta. The project has been plagued by enormous time and cost overruns, poor project management, split responsibility, squabbling subcontractors, and poor coordination. All in all, this chapter offers various problems to be studied and lessons to be learned.

Authors' Note: Just as we were putting the finishing touches to this chapter, we received news that the Calcutta Metro is now fully commissioned—at last! Extracts from this landmark news item include the following (Mukherjee, 1995: p. A1):

> *It has taken twenty-two years to construct 16.26 km [at a] cost of Rs 1,600 Crore. Having patiently borne the dismemberment, the desecration, . . . noise, [and] inconvenience . . . the citizens . . . now see an end to their suffering. Using the world's most labor intensive technology [and a] rate of progress of 0.74 kms a year, . . . getting funds was a problem.*

Just so the reader can understand the magnitude of the project, the total cost noted above is about $5 billion, as one U.S. dollar is approximately 34 rupees. The project thus qualifies as a mega project in the same category as the Eurotunnel completed a year earlier at a cost of $16 billion.

THE STORY IN A NUTSHELL

It is amazing how this mega project started from a trivial incident. India's three largest cities were vying with each other for a government-financed metro rail project. In 1972, the West Bengal Chief Minister, Siddhartha Shankar Ray, had a brainstorm on the eve of Prime Minister Indira Gandhi's visit to Calcutta. He ordered a long stretch of ground around the vicinity of the proposed metro site to be dug up, leaving huge dirt mounds along the Chowringhee road. In reply to Gandhi's query about the reason for this activity, Ray answered that work on the underground metro had already started! The ploy worked. The Central government stepped in, with Indian Railways taking over the entire responsibility for the project, including shouldering its financial burden. In this way, Ray was able to take a West Bengal state project and turn it into a national one (Laha, 1994).

By April 1988, less than half (7.5 kilometers of the planned 16.3 kilometer distance) the total stretch had been completed. In October 1994, another 35 percent of the line was finished (5.7 kilometers) and in September 1995, the balance was finished, two months ahead of the "re-re-revised" schedule. Commuters who had suffered agonies all these years heaved a collective sigh of relief. Nevertheless, the project seems still to teeter on the brink of financial ruin, with the Metro's operating expenses running at 120 million rupees while it is generating less than one third of that total in revenues. With the entire system now operating, the number of commuters expected to use the system each month is estimated at between one and one and a half million. That estimate may be over-optimistic. As one critic noted back in 1994, the huge investment in the Metro would have been sufficient to pay for cars for each of the rail line's passengers.

Beset with losses, the Calcutta Metro continues to hit rough patches. Following the Metro's completion in 1995, twenty-three years after groundbreaking, the first complete trip along the 16.45 kilometer route was completed in just thirty-three minutes. Within days, thirty-eight passengers were hospitalized from smoke inhalation due to electrical fires in the coaches. Meanwhile the losses continued to mount. For every 100 rupees the rail line takes in, its expenses total 278. Breakeven passenger use is estimated at 500,000 riders a day. The line currently services less than one quarter of that total. To make up the shortfall, the Metro has tripled its fares, further exacerbating the problems of attracting riders (Banerjee, 1995). What is the chain of events that led to the development and eventual completion of this enormous project?

A True "Mega" Project

Calcutta, the "City of Joys," has truly earned the pride of having the first underground railway in India. It stands as a landmark, and marks a great engineering milestone in the urban transport history of India. The planning and execution of this project and successful operation of the first phase is a tribute to the indigenous engineering skills of the highest order.

Why develop a metro system at all? The roads in the metropolis constitute a mere 6 percent of the urbanized area, far below the minimum 30 percent required for most metro systems to operate efficiently. Even the meager city roads and highways have been deteriorating, and vendors have intruded onto pavements and parts of the roadways. As a result, the city transport is not only slow, but also dangerous. On a typical weekday, the buses and tramways were coping with some 5 million passengers over this scanty road and tramway network—with unimaginable overcrowding.

Calcutta's most ambitious project, the Metro Railway, gives the city a clear distinction in the country. Other metropolitan cities, such as Bombay, Bangalore, and New Delhi, still talk of developing a mass rapid transport system. But the Calcutta Metro is up and operating. As we describe later, the metro was opened in stages, initially 4 kms (from Esplanade to Bhowanipur) in 1984; a 4 km stretch (Bhowanipur to Tollygunj in the south) in 1986; and finally, 6 km in two stretches (from Esplanade to Dum Dum in the north).

A small, multicolored, folded brochure was issued on the occasion of commencement of commercial service in two stretches between Esplanade and Bhowanipur (October 24, 1984) in the south and between Dum Dum and Belgachia (November 12, 1984) in the north. It lists the metro's salient features:

Total route length	16.43 kms
Stations	17, 15 underground, 2 surface
Coaches per train	8
Maximum speed/average	80/30 MPH
Voltage	750 DC
Total power required	53 megawatts (65 MVA)
Each coach to carry	54 sitting, 218 standing, 2500/train
Interval between trains	2 minutes (at full operation)
Environmental control	ventilation with cool air

With some 65 percent of the work done, project completion at that time was targeted for 1990 at a total cost of 8.6 billion rupees. As noted earlier, when actually finished in 1995, the project's total cost was 16 billion rupees. It seems that the Metro and another major project, the second Hooghly Bridge, also in Calcutta, had been vying with each other for many years to be the world's slowest implemented public projects. The Metro has emerged as the winner, as the bridge was finally completed a year earlier.

CHECKERED HISTORY

The chronic and burgeoning transportation problem of Calcutta had been crying out for an urgent solution for a long time. Exploration of a rapid transport system in Calcutta started as far back as 1947, when the highly popular chief minister of West Bengal, Dr. B.C. Roy, took the lead in shaping a comprehensive plan for city mass transportation, initially imagined as a circular railway. However, its potential use was found to be rather limited; estimates projected that at most, 10 percent of the total population would make use of such a configuration. A detailed study by a British consultant in 1964 led to a proposal for an elevated railway in two sections: A 10 km north-south orientation and a 3 km east-west, with a carrying capacity of 25,000 passengers per hour. An underground system was ruled out at the time on the basis that it might be used as a shelter for the pavement dwellers. Further, an underground system was seen as very expensive, requiring large foreign loans and a heavily subsidized fare structure that would raise questions about the Metro ever becoming profitable in its own right. A somewhat similar recommendation to create an elevated railway was made by experts from the London Transport Board, and later by experts from the French Metro. The final plan was given a concrete shape in 1967. The formal sanction came five years later on the eve of a foundation stone laying ceremony by then Prime Minister, Indira Gandhi on December 29, 1972, and the nominal construction work started the following year.

We use the term "nominal" purposely, because nothing substantive happened until six years later (1978-9) when the yearly fund allocation exceeded the critical mass of 100 million rupees. As we will see below, the funding schedule was such that little could be accomplished prior to the year 1978. Then too, a most unconventional and undesirable start added considerably to the cost and implementation problems. Metros worldwide are usually started from the two terminus stations (surface) and work their way down to the underground section gradually. In Calcutta, due to non-availability of the terminals, initial work was entirely underground for the central sector. This posed special problems, because the central section was unapproachable by either road or rail. As a result, while work proceeded underground somewhere in the middle of the track, it was impossible to bring Metro coaches into the tunnels. Further, the difficult locations made maintenance work time-consuming. Finally, commissioning the central section without either terminal station on line posed unique logistical problems (Sengupta, 1985).

The work started haphazardly, thanks to the difficult logistics. Added to these problems was the primary consideration of not disrupting the normal commercial life and activity of the highly crowded metropolis. The work started simultaneously at two places in the middle of the route. Finally, a totally unrealistic and unpractical target was set: To start operating the Esplanade to the Bhawanipur section (3.5 km) in the middle of the route by the end of 1984. This directive came at the insistence of then railway minister, Ghani Khan Choudhury. In spite of the infeasibility of completing the necessary work within this time schedule, work was directed toward its accomplishment, at the expense of every other duty along the project. The result was a substantial cost overrun. It became clear that the minister's directive was impossible, since this tiny stretch of the rail line did not open, in spite of crashing all activities, until two years later (late 1986). This strategy was widely seen as more of a political expedient that did little except contribute greatly to both the overall cost and schedule slippages for the entire project. In retrospect, due to this short-sighted policy, the "Pride of Calcutta" proved to be Calcutta Metro's mortification (Johari, 1991).

THE END OF THE BEGINNING

Finding the money to support this project has been a challenge from the very beginning. Though the Metro project was approved in 1972, adequate funds could not be provided for almost seven years. A major project of this size needs a certain minimum funding level to make any headway and this was not forthcoming until 1978, thanks to the efforts of then chief engineer, G.N. Phadke. The allocation started with a miserly 61 million rupees budgeted (about $2 million) in 1973–74 and was only increased to 50–90 million in 1977–78. Considering the size of the project, 100 million rupees was considered the minimum critical funding necessary to make a substantive start. This funding level has only occurred since 1978–9, and a steady increase in this figure has enabled the constructors to keep up the tempo of activity at this project. The practice of doling out funding with an "eye dropper" remained a feature of the project throughout most of its life, thereby ensuring that progress would continue at a snail's pace. Consider the following table, showing the dates and amount of funding allocated for this project:

Date	Funding (all figures in millions of Rupees)
1978–9	140
1979–80	192
1980–1	273
1981–2	315
1982–3	435
1983–4	640
1984–5	705
1985–6	772
1986–7	840
1987–8	890
1988–9	1060
1989-90	1190

Clearly, the lack of availability of adequate funds was one important problem with expediting the project. It was, however, by no means the only limitation, as there were many constraints on the project:

- Exclusive possession of the site was not possible—normal life in the city had to go on as usual. While this is a normal by-product of most metropolitan construction, in a city the size of and as congested as Calcutta, with an inadequate road system to begin with, there was a constant battle between maintaining normal traffic flows and creating enough buffers to allow for Metro construction.

- Traffic could not be fully diverted from the roads. As mentioned, the road system was almost completely inadequate to start. It was therefore impossible to divert traffic from already dreadfully overcrowded thoroughfares.

- Impracticability of site protection—the Metro construction remained an open show for all citizens. It also made for dangerous working conditions, not only for the workers, but for passersby.

- Uncharted utilities (sewer, water, gas mains, phone, electric cables). By far one of the worst features of construction was the near total absence of a master plan showing the location of underground utilities and cable lines. Work was continually started, stopped, and gradually restarted as crews hit underground lines and had to call for repair crews and await their repairs to the infrastructure.

- Shops around the site needed to be provided with a view and an approach. Local shop owners were justifiably worried about the effect on their commerce of closing off access to their shops while construction went on. They actively worked against Metro development until they were individually provided with access lanes from their shops to the still-operating streets.

Under any circumstances, the above list would have led to near-impossible challenges for the construction crews. Add to this the usual and difficult problems encountered for underground Metro construction compared to normal open line railways, and we begin to get a sense of the financial and technical challenges facing the project from its inception. Newspaper and magazine headings like "Metro Magic Wearing Off?" and "Another Delay," (*Economic Times*, 1988; Gupta, 1990) became quite common and expected. The cause of the second headline announcing another delay was the interunion rivalry among groups working on the Metro, common in Calcutta.

The project experienced severe flak from the Public Accounts Committee (a sort of watchdog group for the government) for the severe time and cost overruns, as well as its highly unprofitable operations. The number of commuters on the first completed stretch turned out to be a mere 10 percent of the original estimates, with an annual operating loss of Rs 47 million. Even at peak hours, the traffic is unlikely to be more than 50 percent of capacity. As a result, the project has not been able to recoup original spending outlays, although proponents had consistently pointed to the profits the metro was expected to generate. This puts the Calcutta government in the awkward position of having overspent its original project estimates by a factor of ten and having no viable means of reaping a reasonable payback.

THE PROJECT PLODS ALONG

With the opening of the second phase of 5 km from Tollyganj to Rabindra Sadan on April 28, 1986, the mood in the Metro city was distinctly upbeat as real estate prices all along the completed section began shooting up. The Metro's general manager, H.S. Sinha, had reasons to be cheerful, noting (Banerji, 1986):

> *Initially almost everybody took the Metro with a pinch of salt but today Calcuttans are actually realizing the tremendous social benefits it is already generating.*

Supporters of the Metro have always believed that it represented a generational leap for a city with obsolete tramcars, primitive cycles, and hand-pulled rickshaws, and choked by enormous crowds. A simple, five km journey through the city takes nearly an hour by bus, whereas the Metro covers the ground in just fifteen minutes. When the full 17 km Metro stretch is completed, it is expected to carry 1.7 million passengers daily, three times more than the most heavily used system in Mexico City. The present project, completion of the original 17 km route, is only part of the master scheme of the three lines, totaling 60 km. The estimated cost of the first line of the 17 km stretch, i.e., $40 million (at the then prevailing exchange rate) per km is said to compare favorably with metros in the third world, e.g., Sao Paulo, Santiago, and Mexico City. Unfortunately, with the city-state tussle at the political level (in the city, the Congress party is in power, whereas in the West Bengal State, the Marxist Communist Party is in power), the state government loses no opportunity to deride the waste associated with the Metro, casting the project into further disrepute in the city and across the country.

Characteristics of the Calcutta Metro

The Calcutta Metro project was conceived, planned, designed, and executed by the Metro Railway Organization of the Indian Railways, using an entirely Indian workforce, technology, and resources—a wise decision in order to achieve self-sufficiency in this area. The experience gained was intended to be put into use in similar projects in other Indian metros. Naturally, being the first project of its type in India, several problems arose, and mistakes were committed, but hopefully, lessons were learned. These problems and lessons will be discussed in more detail below. The unforeseen behavior of soil or sub-soil water led to costly delays and painful lessons that can be applied to other metro efforts in the future. Fortunately, just about every conceivable detail was well documented by the principles involved. This is literally a "Bible" for metro railways. It has five major parts:

1) History of the project, travel statistics, geology, and climate
2) Civil engineering, mainly cut-and-cover method

3) Rolling stock and the electrical installations

4) Signaling/telecoms, train radio, and computer control circuits

5) General administration and operation

In size, the Calcutta Metro ranks far down the list of metro railways, led by London (270 route-miles, started 1863), New York (230, 1868), and Moscow (225, 1935). But in terms of population and the requirements of a rapid transit system, it had qualified as a world-class system a long time ago. This is very clear from the following extract regarding the global trend for metro construction (Metropolitan Railway Chronology, 1991):

> *After a fairly constant rate of new construction—two systems over five years between 1907 and 1935, the pace of new projects stepped up sixfold . . . Of the seventy-one Metro systems existing around the world, sixteen new systems are now under construction with a total of 430 km and twelve other systems totaling 250 km are in the design or planning stage . . . How big a city . . . to qualify for a Metro? [O]ne million has generally been regarded as a suitable threshold.*

An analysis of prevailing traffic revealed two principal corridors—one along the north-south traffic and the other along the east-west corridor. The passenger flow is tidal and in both cases, it converges into the central business area. As expected, there was a pronounced (85,000 passengers per hour) morning peak (9 A.M.–11 A.M.) and a less pronounced (77,000 passengers per hour) evening peak (4–6 P.M.). The maximum transit movements occurred in the Calcutta central area. The first priority was found to be a north-south axis from Dum Dum to Tollygunj. Finally a double rail system was adopted, mostly in the underground formation.

The keynote style for station architecture was to be simplicity of design, durability, and easy maintenance. In view of the anticipated extremely heavy traffic density, the materials selected were designed to be very tough and durable. To bear the heaviest brunt of the traffic, flooring is made out of very hard and durable stone (Kota stone from Rajasthan) of a pleasing color that will take a reasonably good polish. To mitigate the monotony of a single color scheme, white marble strips have been used both longitudinally and across. The circular pillars along the central line of platforms and in the external face of the staircase have been lined with colored glass, ceramic tiles, and marble strips in different colors to give each station an identity and a clean, pleasing appearance.

Station platforms have been kept totally free from any obstruction, except the essential staircases and pillar supports. At the end of the platforms, cubicles are provided for the operating staff. At the four corners of the platforms, narrow staircases allow the maintenance staff to access the track level. The overall facilities cater to a peak traffic of 3,500 passengers per hour per square yard. Passages, corridors, and subways are designed for 5,000 passengers per hour per square yard for one way movement and 4,000 for two-way movement. Adequate car parking space was provided at heavy traffic stations like Esplanade by widening the roads on the surface.

FINAL PROJECT COST

On the matter of total project cost, a news item about a year and a half earlier had this to say (*Business Standard*, 1994: p. 7):

> Original estimate (1971): 1.4 billion rupees
> Final expected, (end 1995): 16 billion rupees

Overall, the project overspent its original estimates by more than a factor of ten, propelled in no small measure by the substantial upward budget revisions made almost every year. Likewise, the project has consumed over twenty-three years from its inception to completion. The budget for 1995–96 provided 1.4 billion rupees to help complete the project by December 1995. Mercifully, its actual completion had taken place a couple of months earlier. The article also added that while the Metro still awaited 100 percent completion, the cumulative operating losses for the already commissioned stretch of about 10 km by March 1995 were about 4.8 billion rupees, a picture depressingly similar to the Channel tunnel.

A scheme designed to reduce the losses has recently been introduced by Calcutta Metro, in a case of better late than never. By mid 1995, Metro authorities were entrusting the upkeep of some of the stations to a Delhi-based advertising firm, in exchange for exclusive advertising rights. These have proved extremely remunerative in the other metro cities, as well as in metros around the world (Madhavai, Bhattacharya, and Athale, 1995). The trick lies in exploiting the advertising value of the prime real estate and converting it into a true commercial proposition.

Bombay's newly completed and inaugurated Konkan Railway stations (surface, not metro) have been turned into multistoried commercial complexes, built entirely from the funds of participating companies. Such a scheme for Calcutta Metro could have largely helped finance the entire project for the benefit of everyone concerned and without any financial burden on the central exchequer. A lesson drawn from the Hong Kong Metro, and being put into practice at New Bombay with great success will be discussed later. But this model has to be built into the project right from inception.

PROBLEMS UNCOVERED AND LESSONS LEARNED

Overall, the problems of the railways are rooted in two principle sources: Failures of project management and poor economic planning. The Railway Ministry's self-congratulatory references to their so-called "innovative marketing" and improved operating efficiency have been generally seen as eye-wash by the public. Instead, the facts suggest that far too much time and money was being spent in overhead costs not directly related to construction. For example, the wages, allowances, and pensions in the entire railway project absorbed over 40 percent of it's annual expenditure. This figure, however, is merely symptomatic of greater levels of inefficiency, as governmental controls continued to threaten the Metro's ability to deliver on its promises. Employing a massive workforce, using land that was forcibly acquired from citizens at artificially set, rock bottom prices, and continuing to maintain a vast welfare network which included free housing, hospitals, and holiday homes for employees, constituted a virtual state within the state. The huge internal operating bureaucracy added thousands of additional employees to payrolls and continued the Metro's record of heavy losses, without adding much value. The end result was that additional governmental funding for the project went predominantly to support the existing social and bureaucratic organization, with little of the money actually leading to excavation, construction, and other direct project costs (Thakore, 1994). Further, because of the entrenched social programs and safety nets for project workers, there was little incentive to actually finish the project.

The problems that developed in constructing the Metro can be grouped together in three broad categories: General Problems, Technical and Construction, and Contract Difficulties. Each of these classes includes several specific problem areas as well as some potential solutions and lessons that were learned (Phadke, 1985; Tathagata, 1985).

General Problems:

- Administrative: Overhead costs were staggering due to the enormous social umbrella that the contractors supplied for workers and their families. It created large add-on costs that slowed the process of project development. Further, the large bureaucracy made it impossible for any decisions to be made in a timely manner. Every problem was kicked up and down a long chain of command before consensus emerged.
- Land acquisition: Land along the rail lines was appropriated at very low prices from the owners, as the government set values and gave those unsettled by the process no avenues of appeal. This process created very bad feeling about the project, not just from the dispossessed, but from other citizens as well.

- Legal inadequacies: There were no outlets for any individuals appealing the decisions of the government and contractors in the building operation.
- Traffic diversion: This problem took on nightmare proportions because of the fact that there were few alternative routes for passersby to take when a thoroughfare was closed. Because of the huge downtown population, any traffic diversion led to long-term chaos.
- Sporadic shortages of vital materials: Because of funding inadequacies, the project was forced to operate with minimal stockpiles of raw materials. At times, this situation resulted in shortages of cement, forged steel, stone, and other materials, causing interminable delays and stoppages.

Technical and Construction Problems

- Earth excavation and disposal of surplus earth: The amount of earth moved for this project was huge. This problem was further complicated by the fact that early excavation occurred in the middle of the rail line, with neither terminal point completed. Consequently, it was difficult to get earth moving and excavation equipment into the site, and the logistics of hauling became very cumbersome.
- Settlement of the buildings around sites of deep excavation: A by-product of deep excavation is the likely settling of buildings within the immediate area. The Calcutta Metro had a number of sinkhole problems appear as the tunnelling commenced, slowing down the process as more and more workers were reassigned to repair and maintenance duties.
- Utility diversion problems and their effects: Utility problems were a nightmare from the beginning as it was discovered that there was only a rudimentary knowledge of the location of underground lines for steam, gas, water, and electricity. Several times, work had to be halted while repair crews were sent out to the construction site to repair breaks as the result

of drilling or earth removal. One benefit from the process was a much more thorough charting of these underground lines for the future.

- Flooding during construction stages: Calcutta occupies a low-lying geographic region that makes underground water seepage a problem. Coupled with the annual monsoon, a time of torrential rains, the flooding problems became terrible. The contractors had to constantly monitor water levels and, in some cases, shore up collapses caused by excessive flooding.
- Problems of compressed air shield tunnelling in soft soil: On-site engineers constantly had to make adjustments to the normal tunnelling techniques due to the nature of the soil composition. Soft soil led to "blow outs" from the use of compressed air in the tunnelling process.

Contractual Problems

- Contract awards: Perhaps the biggest problem here had to do with the manner in which construction and other contracts were awarded. The decision criterion was predominantly lowest bid. As a result, many bidders who won contracts found they had to cut quality corners in order to make a profit off their contracted prices. Even with an increasing cry for additional criteria, such as past performance or capability assessment, the exclusive use of low bid for contract awards caused far more overruns in the long term than if they had established more rigorous up-front standards.
- Civil engineers were out of the contract loop: The project's civil engineers had no understanding of contract terms and could not arbitrate or settle claims. When contractors did not satisfy their legal obligations, the civil engineers on site were in no position to solve the problem. Instead, the arbitration procedure went up through multiple levels of administra-

tion and caused long delays until temporary settlements could be reached.

- Arbitration was long and painful: The entire arbitration process took far too long and led to many work stoppages. The project needed to establish a more streamlined process for adjudicating claims from either side.

The Calcutta Metro project offers a number of lessons for similar projects both in the developing and the developed world. The project, as is common with most major projects in India, has suffered due to far too many small subcontracts. With hundreds of subcontractors all providing their incremental added value, it has been extremely difficult to adequately track deliveries and assess performance of the subcontractors. Also, the credentials of many of the contractors and subcontractors are doubtful. Because of political and time constraints, the project's developers were often forced to accept subcontractors without adequate back checks of their reliability or past performance. The main criterion in awarding contracts was simple: Lowest price always wins. This approach was most unfortunate, and in most cases what appeared to be the lowest price may actually end up with a price higher than the highest bidder due to hidden costs and the expense of conducting rework because of material deterioration. This problem was repeated far too often and had serious detrimental effects on the project's schedule and costs.

What was required was a proper evaluation of the tenders, not merely on the basis of quoted price, but particularly its evaluation in the context of the contractor's credibility and past record on similar projects. Such information is not easily available, certainly not published or widely publicized, unless the contractor has been a total failure. But getting a contractor's reference list, and spot checking with some of his clients can go a long way in helping to choose the right contractors. The price really should only be one of many considerations. In fact, some of the best executed projects have been by reputable contractors even without a formal quoted price!

One can also draw upon the experience at similar projects elsewhere in the country or even abroad. As an example relevant to the Calcutta Metro project, the experiences of the MRTC (Mass Transit Railway Corporation, Hong Kong) can be useful. This highly successful structure has developed basically from a project management organization into an operating company. The operations director was appointed long before the first line came into operation. He was fully involved in the design and construction phase, in addition to reviewing and finalizing operational policy, training programs, maintenance plans, levels of equipment and staffing, to ensure smooth and efficient operation right from the start.

The engineering group looked after all the front end activities of work planning and feasibility, civil engineering design, architecture, permanent geotechnics, and materials and quality control. The project group had the specific task of day-to-day supervision and some of its senior managers were delegated the specific powers of the engineer.

One of MRTC's first tasks was to build an effective project management team with members from diverse disciplines and experience. The corporate core values, having been laid down and followed (built to standard, within time and budget), have proved to be of immense benefit. As part of this process, strict and formal financial controls were introduced right at the inception stage. Programming methods used within the British nuclear industry were introduced and network analysis was employed throughout the project execution. Finally, the Hong Kong government provided full support to the project with minimal interference or direction. Rather than making the metro development a political issue, as in the case of Calcutta, the Hong Kong government supported it as a social benefit without excessive oversight.

Of the total cost of HK $25 billion (U.S. $3.2 billion dollars), twenty was spent for construction and five for capitalized interest. A significant part of this funding came from the development of property on the basis that MRTC increased land values along its route, and hence was entitled to a part of the benefit (Sharpe, 1985).

There is no guaranteed formula for success in implementing massive capital works in an urban environment. Each situation needs special consideration. The MRTC's approach has been totally appropriate to the Hong Kong situation, creating a team effort to achieve a major objective.

FUTURE METRO PROJECTS

It is clear that no transport system can match the Calcutta Metro when it comes to moving large numbers of people quickly. But its construction and operation required massive investment that was, in the final analysis, poorly administered. The project was handcuffed for years because government officials were unwilling to provide enough funding to make more than token progress on the rail line. In a developing country with limited resources, it was necessary to spread this investment over many years. Nevertheless, with the accompanying inflation and escalation of costs, the end result was only to push off the completion and add huge amounts of money to the overall cost (Phadke, 1986). This problem was particularly sharpened due to the huge social net that the project provided, consuming 40 percent of all funding in overhead expenses.

With regard to financing, it is clear that a large scale public transport scheme, especially the expensive Metro, is bound to make a loss on its direct cost. The beneficiaries include employers, employees, offices, shops, places of entertainment, schools, hospitals, and property owners. Affordable fares have to be lower than their real cost. Dr. Tony Ridley, the creator of the Hong Kong Mass Rapid Transit System, argues for a theory of "Benefit Sharing" in which third parties who benefit must also contribute. But this can happen only after the project is implemented. In practice, property owners and developers have turned out to be the most fertile sources of funding in the absence of adequate government support.

As soon as a project is announced, the land values in the immediate vicinity of the Metro will increase following the mere rumor of construction. Ridley argues that municipalities should acquire these extra revenues through taxes on specific beneficiaries. The spare spaces between the stations and the tracks and within the station concourses can also be tapped in order to meet the operating and maintenance costs.

This revenue development pattern is being used in the New Bombay area beyond Thane Creek Bridge—five new stations serve the extension of the VT-Mankhurd line beyond the Thane Creek. These have been designed as multistoried commercial complexes, only the ground floor being used for tracks and platforms. Commercial areas are leased out for periods of sixty years, and from the proceeds, the developing agency is not only funding two-thirds of the cost of the line and the bridge across Thane Creek, but is retaining some surplus funds. Thus, even in India, dependence on government for funding can be minimized through proper planning and advance action (Phadke, 1990).

References

Banerji, I. 1986. "Calcutta Metro Moves," *India Today,* May 15, p. 107.

Banerji, R. 1995. "Metro Rail—Running into Hitches," *India Today,* Oct. 31, p. 197.

Business Standard. 1994. "Metro Rail to be completed by 1995 end," Feb 5, p. 7.

Economic Times. 1988. "Metro Magic Wearing Off?," Dec. 23, p. 1.

Gupta, S. 1990. "Another Delay," *Business World,* Dec. 6, p. 14.

Johari, H.C. 1991. "Metro Railway—The Pride of Calcutta," *Economic Times,* December 19, West Bengal Supplement, p. 10.

Laha, A. 1994. "A Giant Leap underground," *Business World,* June 1, pp. 150–152.

Madhavai, M. S. Bhattacharya and Guri, A.A. 1995. "For Clean Image," *Financial Express*, August 22, Habitat Supplement, p. 1.

Metropolitan Railway Chronology. 1991.

Mukherjee, S. 1995. "Calcutta Metro rail line will be fully commissioned today," *Times of India*, September 27, p. 1.

Phadke, G N. 1985. "Constraints and problems in the Construction of Metro Railway, Calcutta with Special Emphasis on the Engineering Aspects," presented at the International Seminar, Calcutta, November.

Phadke, G.N. 1986. "Metro Pitfalls," *International Railway Journal*, May, pp. 37–38.

Phadke, G.N. 1990. "Development of Urban Transport in India," Key Note Address at the Conference organized by RITES/Ministry of Urban Development at New Delhi, June.

Sengupta, S 1985. "Problems regarding Design, Construction and Maintenance of Metro Rolling Stock," presented at the International Seminar, Calcutta, November.

Sharpe, D.J. 1985. "The Experience of the Hong Kong MTRC," presented at the International Seminar, Calcutta, November.

Tathagata, R. 1985. "Contracts in Metro Construction: Some Observations," presented at the International Seminar, Calcutta, November.

Thakore, D. 1994. "Why Indian Railways is Running Out of Steam," *Independent*, March 25, p. 8.

10

The Iran Petrochemicals Project

IT IS WELL-KNOWN that oil and water do not mix, nor (for that matter) do petrochemicals and water. Here we bring you the classic case of the $5 billion Bandar Shahpur petrochemical complex in southern Iran, an Iranian joint venture with a consortium of Japanese companies led by one of their foremost corporations, Mitsui. The Bandar Shahpur project is a classic example of the problems that can occur in joint ventures between foreign companies and local governments. In this case, the story is also heavily laden with examples of political unrest, revolution, and war, and an incremental, on-again, off-again approach to project development. But first a bit of the background on the Iranian petrochemical plant.

A DESERT DREAM

Back in 1968, Sueyuki Wakasugi, a top official in the Japanese Mitsui group, was driving through the hot barren wasteland near Abadan, some 700 km south of Tehran. He saw something quite picturesque: The gas flaring in the distant oilfields. That sight gave Wakasugi the idea to convert Iran's wasting resource, natural gas, into profitable petrochemicals through the use of Mitsui technology. Unfortunately, his dream led to the birth of one of the most expensive and the most embarrassing investments ever undertaken by a Japanese company (Wiegner, 1979).

In 1970, the project made technical, commercial, and even political sense. A waste resource was to be converted into valuable products for which there was both a direct need within Iran and a growing world market. Further, codevelopment of the project with the Iranian government would help cement Japanese ties in the region, so critical for an economy that was forced to import 100 percent of its oil. The final project agreement consisted of a world scale ethylene cracker, with a capacity of 330,000 tons per annum of ethylene, together with other derivatives and liquefied petroleum gas. A 50/50 joint venture company was formed between various Iranian and Japanese companies.

THE DREAM IS SHATTERED

An integrated petrochemical complex typically centers around a cracker to produce olefins (ethylene, propylene etc.). Nearly half of the production (capacity of 330,000 tons per year), was earmarked for export and the rest for downstream products (polyethylene, etc.) for consumption in the growing domestic market. Iran, with its large population, has a strong market for petrochemical products, unlike the other oil producing countries of the Middle East. The petrochemical capacity/production in most of these countries is targeted to foreign markets. The idea is to add substantial value to the better-known exports in the form of crude oil. So far, so good but there is an important catch for these Middle Eastern countries: The technology necessary for this refining process is extremely high-tech and has to be imported from the West or from Japan. The products are likewise expected to be exported to the same areas and/or the established markets of the technology supplier. Naturally, this produces a clash of interests, as those possessing the technology expect to be the first to exploit the natural resources while the Middle Eastern countries, although not equipped with the technology, want to maintain their prerogatives in regard to export and sale.

Saudi Arabia was probably the wisest among the oil producing countries in terms of having a sensible policy in this regard. Their technology agreements with Western companies are often tied up not only for foreign assistance in production, but also in sales and marketing. In many cases, they even welcomed financial participation from the very same partners. This stake ensured the collaborators full support in all of the activities relating to production facilities. Unfortunately, this approach was not following in the example of other oil producing countries in the Middle East, including Iran.

With the wave of nationalization in Iran following the Iranian Revolution, four of the smaller American and Japanese joint ventures there were the proposed targets of takeover by "mutual consent" by the Iranian National Petrochemical Company. However, the Japanese joint venture, Iran-Japan Petrochemical Complex, appeared as a too large and too complex project for the proposed takeover. Moreover, the project was unfinished. For this 50/50 partnership, Iran did stake her claim for management control in the spirit of nationalization, making the situation ever more complex.

A short history of the project reads like a bad adventure novel: The original contracts for the plant were negotiated with the Shah's government. In the midst of project development in 1978, came the Iranian revolution and all work at the site ceased. The original estimate for the project had been some $500 million, but by 1979, with the revolution over, the estimated cost had risen to some $1 billion. In spite of the cost escalation, work on the project was resumed as Mitsui continued to pour in money. But work was again brought to a sudden stop by the outbreak of war between Iran and Iraq in 1980. By that time, at least $3.6 billion had already been spent and it was estimated that a further $2 billion would be required to get the plant repaired, rehabilitated, and working properly (*Economist*, 1981).

Mitsui was faced with a true dilemma. They had already invested over $2 billion of their own money in the venture, and they were now faced with the equally unpalatable prospects of funding the project through to development or pulling out and writing off a huge investment. How they arrived at that state presents a fascinating story in international business relations coupled with an inability to correct the tide of events that continued to sweep everyone, particularly Mitsui, along for a multibillion dollar ride.

The political repercussions of this project were evident thousands of miles away in Japan as well. Mitsui's stake in the project was substantial enough for them to persuade the Japanese government to declare the IPP as a "national" project. Apart from its obvious benefits, it entitled Mitsui to the low-interest foreign aid loans required to complete the project. Despite such support, uncertainty and delay continued to plague this major project. Somewhat frustrated with such developments, it is no wonder that a senior officer of Mitsui is reported to have said:

> *Our abacus tells us that we should have pulled out of this mess a long time ago.*

But, of course, it is easier said than done.

A Chronology of Disaster

One way to analyze the flow of this project (note we specifically do not use the term "progress"!) is to follow some of the principal milestones going back to its original inception and following the major developments through to its current state.

In January 1979, a consortium of fifty-four Japanese companies looked into the feasibility of developing a large petrochemicals manufacturing facility in northern Saudi Arabia. Later that month, after securing $158 million dollars in advance financing from a syndicate of twenty of Japan's largest banks, Mitsui decided to develop, in partnership with Iran, a massive petrochemical complex in the Abadan region. In October of the same year, Mitsui won governmental support and their supplemental investment of $88 million in the joint venture.

Iran's willingness to work with Japan had more than simple financial motives propelling it. The United States had called for heavy sanctions against Iran in the wake of their seizing the U.S. embassy in Teheran. Japan publicly opposed the American aims. By early 1980, Mitsui had invested $1.5 billion in the Abadan complex and the Japanese government, fearful of Iranian retaliation, portrayed itself as neutral in the conflict between the two countries. In February of 1980, Mitsui had secured an additional $83 million commitment from its government. The cost of the plant, expected to be $2.3 billion, had risen in just one year to an estimated $3 billion.

Despite the international outcry and partial embargo of Iranian oil exports, the Japanese agreed to resume work on the (now) $3.2 billion facility in May. Work proceeded quickly as Mitsui poured in an army of technical and logistical personnel to speed production. All indications were that the plant would come on line by the end of the year, following Mitsui's crashed schedule and huge monetary expenses.

In September of 1980, the two largest armies in the Middle East clashed as Iraq took advantage of Iranian political turmoil to invade and attempt to annex the strategic Basra-Abadan region. On September 25, Iraqi warplanes bombed the soon-to-open plant, causing minor physical damage but giving Mitsui a huge psychological scare. Two weeks later (October 9), the plant was bombed for a second time. Again, damage was light but the plant lay in the midst of an Iraqi army push to seize the area. The Japanese, who had already invested $1.4 billion of their own money in the project had to decide their next move. The Iranian government had assured them that with an additional $60 million, the facility could begin operation.

Mitsui, senior partner with four other Japanese companies, finally drew the line as it became increasingly apparent that Iran and Iraq were engaged in an all-out war along the southern frontier, in full sight of the petrochemical plant. Consequently, on April 24, 1981, Mitsui announced that pending further negotiations with Iran, they would not invest additional money in the project. Later that same year, the company announced that they would continue to offer technical assistance and were willing to sell their on-site equipment to the Iranian government in preparation for a complete withdrawal. The Iranians, desperate for additional revenues to fund their ongoing war, offered new concessions to the Japanese in return for their promise to revive the project. In late July, following three days of talks with the Iranians, Mitsui announced that they were unable to reach a substantive agreement that would renew the joint venture. By fall of 1981, it appeared all but certain that the Japanese would take their billion dollar loss and walk away. This position was solidified in September, when the Japanese government confirmed their approval of Mitsui's withdrawal from the venture, in effect writing off their own several hundred million dollar investment.

The next eighteen months were spent in a sort of holding pattern as Mitsui held out the possibility of reopening the venture sometime in the future, provided the political and military situation had stabilized and the Iranians were willing to put up more of their own money to offset costs. At the same time, Mitsui negotiated a one-year delay in repayment of multimillion dollar loans to the Japanese bank syndicate.

By May 1983, the Iranian government, although exhausted by the cost of their war in both dollars and human lives, finally gained the upper hand and managed to push Iraqi soldiers away from the petrochemical complex site. At the same time, Mitsui negotiators, anxious to recoup some of their losses in the original investment, agreed on a deal to resume construction of the facility, which had been seriously damaged after almost three years of war and neglect. Iran offered to finance the balance of the cost, which had risen to an estimated $4 billion, in exchange for Japanese redevelopment of the project. Work began again on a complex that was in seriously poor condition. For example, tools had been left scattered about, sensitive equipment was never maintained or even cannibalized, and piping was missing or badly damaged.

In September, as the war with Iraq was winding down, their warplanes staged a last raid over the complex. There were no reports of fatalities and overall damage was considered slight. Nevertheless, the effects on Japanese commitment to the project, by that time wholly uncertain, became even less clear. Work on the plant moved at a snail-like pace as the Japanese were wary of overcommitting themselves to the project a second time. Large sections of the plant were considered to be beyond immediate repair and had to be razed and rebuilt. So the project plodded over the next several years as costs continued to rise alarmingly and the plant appeared no closer to coming on line. By late 1988, Mitsui announced that they were considering abandoning the $4.5 billion dollar plant.

Turmoil between Japan and Iran ensued as the Iranians reaffirmed their commitment to completing the complex despite the disruptions of the Gulf War and their own exhausted finances. Further, international analysts seriously questioned the need for the plant, given the overcapacity with existing facilities worldwide. As Mitsui made it increasingly clear that they had had enough of what they were beginning to suspect to be an enormous white elephant, the Iranian government sued them for $1.5 billion in compensatory damages for walking away from the construction project. In October 1989, following a month of negotiations, Mitsui agreed to pay approximately $1 billion in compensation to the Iranian government for cancelling the joint petrochemicals complex. Though finally free and clear of the Bandar Shah plant, Mitsui's total loss on the project has been variously estimated at between $2 and 3 billion.

Ironically, institutional memories are short, even in the wake of this disaster. By September 1995 the *New York Times* reported that two international companies, Technipetrol of Italy and K.T. International, had agreed to jointly build part of a new petrochemical plant in central Iran. This represents either an extreme case of wishful thinking in light of recent history, or an opportunity for these organizations to learn from Mitsui's errors.

WHY DID THE JAPANESE PERSIST FOR SO LONG?

Probably the most persistent question regarding the entire project mess was why Mitsui stayed with it as long as they did. Through skyrocketing costs, political instability, wars, and even missile attacks, the Japanese conglomerate maintained its allegiance to the government and their commitment to the project. Perhaps the major reason for their willingness to stick out the project was the Japanese government's dire need for the crude oil entitlement they received as a partner in the joint venture. Once the construction work was in full swing, the crude oil entitlement was increased by some 50 percent, to 30,000 barrels per day, as an added incentive to complete the project. And for Japan, energy and oil is a matter of life and death since the country has hardly any indigenous resources. The entitlement came at the time of the oil crisis and embargo was the main attraction for the Japanese corporation's participation in the first place, leading also to the full backing by their government.

Because of the Japanese need for oil, it became essential for Mitsui to try to honor their commitment. Add to this their obsession for "face saving" at (almost) any cost! It is no wonder that despite the enormous cost overruns and serious interruptions and damages caused by the Iran-Iraq war, Mitsui did not give up easily. In this effort, their commitment was matched by the equally tenacious support of the national government and other resources, as they were fully backed by the Japanese consortium of banks and by their government's guarantees. Their willingness to continue to invest in the project is particularly surprising as additional events worked against their interests. For example, with the Iranian partner (the National Iranian Oil Company) failing to provide the money required for project completion pro-rata to its share, the Iranian share in the investment fell below its originally stipulated 50 percent. This alone has serious political repercussions, since the installation is located in the troubled Middle East. It put Mitsui and the Japanese government in an increasingly awkward situation in trying to seek redress and additional investment from the local country.

The project came to its first halt soon after the revolution in Iran. Construction work was resumed one-and-a-half years later, after finalization of an additional financial package from the government and the banks. Later, the Iran-Iraq war threatened the very survival of this strategic project which was vital to Iran. The project site was bombed several times, both from land and from the sea. The full details of the extent of damage to the project have not been reported, for obvious military and political reasons. In any case, the plant and equipment remained idle for over fifteen years in the open marine atmosphere, probably leading to severe corrosion and breakdown. To correct this state of affairs and crash the project's completion would require an enormous Japanese force, perhaps some 6,000 strong, including over 2,000 engineers and technicians. Naturally, the Iranian authorities were insisting that the development make extensive use of Iranian personnel, a politically important but impractical notion, since it would have required extensive training prior to project restart. Finally, there was a considerable "culture gap" between the partners that was made more pronounced by their different strategic objectives. Iran, embroiled in the midst of a bitter war less than one year after a revolution, naturally had different immediate priorities, both for natural resource development and international relationships.

Finally, the costs associated with the project were staggering. Originally, in the late 1960s, the project was estimated to cost $0.6 billion, and this figure had to be successively revised upward in several steps to $2.4 billion, then to $3.5, and perhaps on completion (if it ever gets completed!) to $5 billion and above. No wonder, people in the field of petrochemicals have called the project one of the most expensive white elephants in the history of this industry. Further, as the project "progressed" it continued to attract far more disdain than support. It was suggested in the early 1980s that even with some 65 percent completion, the most sensible thing would have been to abandon the project to avoid sending good money down the drain.

IRANIAN PETROCHEMICALS—WHAT ARE THE LESSONS TO BE LEARNED?

The prime lesson to be learned from this major petrochemicals project (and its disastrous outcome) is that risk analysis can *never* take all political uncertainties into account. Without the political developments, including the Iranian revolution and the Iran-Iraq war, the petrochemical project could well have been completed, albeit late and with a substantial cost overrun. But smooth operation and proper marketing could still have made a success of the project, with benefits to all concerned both in Iran and Japan.

Perhaps the most sensible thing would have been for Japan to call a halt, "pull the plug" in order to cut their losses. But it is easier for an objective observer standing on the sidelines to argue that side than for one of the principals to do so. One keeps hoping that somehow things will turn out to be right at the end. Such is the overoptimistic outlook. Once the project was underway, the options for both countries were rather limited. They both needed to complete the project for their own reasons and consequently, engaged in a cycle of spend, work, and halt while the whole time the clock and cash register were both running.

The Iran project has been labelled as a mixture of oil and water, in this case, "petrochemicals and politics." And yet, the very same ingredients, oil and water, have indeed mixed well in Saudi Arabia. In fact, a much larger project and a fully integrated petrochemicals complex succeeded in Saudi Arabia despite the oil politics. The project offers an interesting contrast to the one seen in Iran. We can briefly consider the reasons for its success.

This short story concerns SABIC, the Saudi Arabian Basic Industries Corporation (Kharbanda, 1989, 1993). With the right strategy of fully partnering itself with foreign companies, not only for oil exploitation, but for a full range of business activities, and keeping the petrochemicals away from politics, the project was singularly successful technically and commercially.

Both these goals (technical and commercial) were assured by choosing the right global partners and ensuring that the tie-up related not only to technology, but also included a significant financial stake. Further, the relationship was understood from the beginning by all parties to extend not only through project execution but also operation and marketing worldwide; in effect, a lifetime "marriage" with full commitment on both sides. Their partners were truly an international mix, including the U.S., Europe, and Japan (the list included a Who's Who in the chemical world, with names like Celanese, Dow, Exxon, Mitsubishi, Mobil, and Texas Eastern), who knew their business and were wise enough not to dump the production indiscriminately in the world markets. Their pricing reflected the global trend and products were carefully marketed so as not to disturb the delicate balance prevailing in the market conditions. You will also notice that ironically, the partners included the Japanese giant Mitsubishi, a close competitor and counterpart of the Mitsui organization that was involved in the Iran Petrochemical disaster.

To bring the story right up to 1995, we see that SABIC features prominently in the global league of chemical and petrochemicals. The prestigious British newsletter, *Chemical Insight,* has an annual feature to rate and rank the major international players in these fields. The ranking is done on the basis of the improvement in the last year's (in this case 1994) results of a company compared with its performance in the previous year (1993), and SABIC features prominently in both. In petrochemicals, SABIC emerged in a "strong second" position after the industry leader Lyondell Petrochemicals. Among the chemical companies, it ranked a close third after Union Carbide and Du Pont. They are literally right at the top of the world in the chemical and petrochemical fields, appearing alongside established giants. Credibility, indeed!

Foreign partnerships certainly can work; we see examples of successful ventures every day. The principle issue that must be addressed as part of any project's risk analysis is the degree to which various forms of trouble are prevalent and predictable. Hindsight is easy and unfair. Certainly, in the aftermath of the Iranian revolution and rise of fundamental Islam there is a tendency for some to assert that they saw it coming all along. Whether or not such claims are true (and more often than not they are false), a far better approach is to expand risk analysis to take into consideration a greater degree of uncertainty. For example, while it is common to assess technical and even administrative forms of risk, the Iran Petrochemicals case teaches us the importance of devoting more effort to understanding potential forms of disaster in foreign markets. Rather than using well-established risk profiles, it is imperative to look at additional factors, including social and political forms of risk. Where large-scale projects are concerned, it is a far safer course to overanalyze the project's risk than to rely on past experiences and hope for the best.

References

The Economist. 1981. "Editorial: 'Mitsui moves to a showdown," Vol. 279, May 2, pp. 79–80.

Kharbanda, O.P. and Stallworthy, E.A. 1986. *Successful Projects With a Moral for Management.* Aldershot, UK: Gower.

Kharbanda, O.P. 1989. "SABIC Comes of Age," *Chemical Weekly,* Nov 21, p. 40.

Kharbanda, O.P. 1993. "SABIC—a desert 'dream' come true," *Chemical Weekly,* Dec 7, pp. 147–8.

Kharbanda, O.P. and Stallworthy, E.A. 1983. *How to Learn From Project Disasters.* Aldershot, UK: Gower.

Wiegner, K.K. 1979. "Saving skin but losing face," *Forbes,* Vol. 124, October 15, pp. 74ff.

Part 3

The Developed World Is Not Immune to Disasters

WE HAVE EXAMINED the failures in the developing world, but the developed world, despite its much enhanced facilities, is not immune. In fact, what often separates the disasters in the developed world from those of its less affluent neighbors is not the size of the failures but rather, their effect on the sponsoring organizations. Because of the case reserves of many of these organizations, they are in a better position to absorb the losses from failed projects and move on. However, there is no guarantee that the organization will be in a position to survive. The huge losses associated with Lockheed's L-1011 project eventually pushed the company entirely out of the commercial aircraft market.

The larger point, however, is that there is no panacea to project failure that developed economies enjoy. In fact, as one noted author writes, it is often precisely the successful and wealthy business environment that breeds that greatest propensity toward future error (Hartley, 1994). Hartley's point is that a history of success is often more dangerous for an organization than one that shows examples of failures as well. Successful companies ultimately begin to show signs of Hartley's "Three Cs" mentality: Complacency, Conceit, and Conservatism. Complacency precludes an organization from acknowledging the need to question its decisions or alter its well established routines. Conceit exists in most successful companies as they assume past successes are indicative of a happy future. Conservatism prevents these companies from taking risks and challenging the status quo. Taken together, Hartley's "Three Cs" have led to more than their share of classic project management blunders.

For example, DuPont's product "Corfam" was a technological breakthrough in the manufacture of substitute leather, and it was hailed as being akin to their development of nylon a few decades earlier. It took nearly twenty-five years of research to perfect the process, yet despite the most careful production and marketing plans, the project was written off as a disaster in the wake of mounting costs and poor public reception and was ultimately abandoned. The leather industry had countered DuPont's claims with a huge advertising budget, portraying Corfam as a cheap substitute for real leather. DuPont gave up on the product and leased the technology overseas. One country in particular, Poland, has acquired the technology and production patents from DuPont. Shoes made from this material are selling well in Poland, and are even being exported to the United States—another example of a development that has been used against its creators.

We mention the above example to illustrate a theme which should be patently obvious to most readers who are currently employed in American corporations: Failure does not respect national borders or past achievement. All organizations have experienced very real examples of project failure. The challenge, as this section's cases will demonstrate, is to understand the nature of the failure and its root causes, and learn the corresponding lessons for future efforts. In fact, the obvious lessons that this section's cases should make clear can be listed as follows:

- Even the most successful organizations make mistakes, but manage to survive because most of their projects are successful.
- Making mistakes can be an effective teaching tool in learning how to avoid similar mistakes.
- Success does not guarantee continued success. Even the best companies are vulnerable.
- Success all too often promotes vulnerability, through complacency and smugness.
- There should be judicious imitation: the operations of successful competitors should be observed and adapted.

Each reader needs to bear these lessons in mind as they go through these examples. How did these organizations get into the problems that they experienced, and more to the point, what can be done to pull them back out? In some cases, the answers may be clear. In other cases, the nature of the failure makes it that much more difficult to assess the root causes of the company's recovery. In all situations, however, these projects represent just a small set of the checkered legacy of project management in developed countries.

References

Hartley, R.F. 1994. *Management Mistakes, 4th ed.* New York: John Wiley.

11

The Sydney Opera House: A Landmark and a Lottery

INTRODUCTION

Even twenty years after completion of the Sydney Opera House, it is still being hailed as a landmark with a wonderful exterior. In spite of the innovative design, or perhaps because of it, its physical properties represent a classic example of unworkability. The interior, though brilliant, is not even suited for opera! To quote from the headlines and beginning of a recent article on the subject (Cargher, 1994, p. 24):

> *"Is half an opera house better than no opera house? Perhaps so, when the half is the major tourist attraction not only of the city but of the whole country in which it is situated . . . No theater in the world has a more spectacular exterior and more beautiful surroundings . . . Yet no opera house in the world is less suited to its name."*

This quote seems to say it all!

THE OTHER TAJ MAHAL

The Other Taj Mahal is the title of a fascinating book by a Sydney journalist (Yeomans, 1968), whose first book *The Scarce Australians* showed him to be a rare combination of first-class reporter and diverting writer. He followed the creation of the Sydney Opera House (SOH) year after year, and soon after finishing the book, the author and his photographer wife left Sydney for a long trip through Europe. We set the scene here by quoting from the book jacket:

> *The Other Taj Mahal is the full and frank story of the most controversial building of our times . . . [a] strange and wonderful white monster . . . Today, more than eleven years after Jorn Utzon of Denmark won . . . the construction is still continuing and so are the arguments . . . Is the Opera House a huge joke? The biggest white elephant in history? Or the greatest architectural achievement of this century?*

Although the finished building is generally acknowledged to be one of grace and beauty, its development was beset by problems from the beginning. So well-known are the tribulations of the SOH's history that it has been featured as a frontispiece in one of the very few books published on the subject of project disasters (Kharbanda and Stallworthy, 1983). With its distinctive "sailing" roofs, perhaps the single feature that most fascinated the judges, SOH is an unforgettable landmark. Though the project has been a miserable failure in terms of project management, its completion has been considered by some to be an example of a "positive" failure. That is, the final product almost (but not quite!) obviates the dreadful mistakes that went into its construction.

The original concept of the SOH was of a design sketch submitted by a Danish architect, Jorn Utzon. It was chosen from an open global competition which attracted over 700 contestants from forty-odd countries. Although lacking the requisite design features, the judges were so much taken in by the sheer beauty of

Utzon's sketches that they labeled his entry as not only the best, but also the cheapest! Wishful thinking?

THE CONCEPT, THE COMPETITION, AND THE AWARD

Early in 1956, the government of New South Wales announced a global competition for the design of an opera house, the relevant conditions for this competition being stated in a twenty-five page booklet. The specifications and the requirements seemed to be crystal clear, attracting considerable interest worldwide. Over 700 contestants paid $20 for the official booklet, but at various stages in the submission process, nearly 500 dropped out. A total of 233 entries were received by the closing date of December 3, 1956; included were entries from Australia (sixty-one entries), Great Britain (fifty-three), West Germany (twenty-six), the United States (twenty-four), Switzerland (six), France (five), Japan (three), and one each Egypt, Ethiopia, Iran, and Singapore. The four judges were distinguished in the field of architecture: one was from the United States, one from England, and two from Australia.

The results of the judges were announced on January 29, 1957. The winning entry was Number 218, submitted by a thirty-eight year old Danish architect, Jorn Utzon. This design was in the exhibition display and many of those present saw in it something transcendental in contrast to the traditionalists who were bewildered and even dismayed by it. Though delighted to win this prestigious global competition, it came as a "great shock" to Utzon. His immediate task was to convert his dream into a workable design and hopefully, in due course, to a reality. Utzon had never single-handedly designed and supervised the construction of any large building from start to finish. But no one could have even imagined the shape of the storms to come.

Structural engineers in Sydney and elsewhere whistled in surprise. Why? Utzon's shell roofs were much bigger and wider than anything ever built anywhere. The engineers were made uncomfortable by what they viewed to be insurmountable technical difficulties. It seems that the judges were so fascinated by Utzon's sketches that they ignored the key fact that there were no plans nor even a perspective drawing for the building as required. In fact, judges anticipating this observed (Yeomans, 1968: p. 34):

> *"The drawings submitted are simple . . . diagrammatic. Nevertheless . . . we are convinced that they present a concept . . . [for one of the] great buildings of the world . . . [It is an] original and controversial design. However, we are absolutely convinced of its merits . . . (despite) many points of detailed criticisms . . . and a number of corrections will have to be made . . . [The] general breadth of the imaginative concept is an overriding consideration."*

What an understatement the word "correction" turned out to be! As events would prove, the building as envisaged (by the winner's sketches) could not be built, period.

After the competition, one story made the rounds: The chief judge, arriving late for the competition, spotted Utzon's entry in the pile of rejected sketches, held it aloft to his colleagues saying, "Here is your Opera House." Both the Sydney judges denied this story, while confirming that Utzon's drawings were already in the "short list," but the chief judge clearly seemed to have been completely sold on the sail-type design. Apparently his colleagues had commented that Utzon's design was marvelous but impracticable. To each of the judges' objections, the chief judge advanced plausible solutions and obviously carried his team with him. His colleagues faced a classic dilemma: gamble and "shoot for the moon," create one of the world's great structural achievements *if* it came off, or opt for the alternative and play safe, selecting one of the conventional designs not posing any unsurmountable problems. Faced with these two options, the judges decided to forge ahead with a dramatic statement.

What did laypeople think of the sketches? Although opinions were sharply divided, the general consensus alternated between the staunchly positive and the derisive. For example, many of the comments varied from "just a bunch of roof" and "Danish pastry," to an "insect with a shell on its back which had crawled out from a log." What about the professionals? The architects, both failed entrants and otherwise, seemed equally split about the judges' choice. Some vehemently approved of Utzon's design, others thought that it was a superb piece of architectural tragedy, possessing a striking look, but totally devoid of interior efficiency. And how did Utzon himself see it? (Yeomans, 1968: p. 45):

> *"If you think of a Gothic church, you are close to what I have been aiming at. Looking at a Gothic church, you never get tired. You will never finish looking at it. When you pass around it, or see it against the sky, it is as if something new goes on all the time . . . with the sun, the light and clouds, it makes a living thing."*

NO DESIGN, NO ESTIMATE, NOT EVEN A GUESSTIMATE

By the time the Sydney Opera House was completed, there was a cost overrun of fifteen times the initial estimates of $7 million (Australian) to a final actual cost of nearly $100 million. But these initial cost figures are essentially meaningless, since at the time of the initial estimate, no design basis existed. In fact, once this entry had been judged as the best, it was not even known whether it could be designed at all as depicted! But more of that later. The project's true challenge most certainly proved to be a developmental one, although never described as such.

The mystery of the SOH lies in the "full steam ahead" mentality which continued to drive its development in light of a lack of fundamental cost estimation based on full schematic diagrams and materials estimates. As such, it is particularly compelling to note that one of the points raised in favor of the winning design was not simply its uniqueness, but also its supposed inexpensive cost. Clearly, there was no basis for estimating (or even "guesstimating") project costs. The perhaps foreseeable result was the advent of runaway project costs and sliding budgets as the project was destined to produce.

In contrast to the common perception that Australians are in general, sports-mad, Sydney also has a vigorous cultural and intellectual life. Hence, the lack of theater and concert halls was strongly felt. A world-class opera house would go far toward readdressing this perceptual shortcoming. A compelling issue was how to finance it. Amidst the controversial and sometimes raucous political climate in New South Wales, someone came up with a brilliant idea. State-run lotteries, which are a very popular pastime, were deemed the logical choice. Rather than invoke additional taxes for its construction, the government used the proceeds of state lotteries, thereby indirectly funding the building with public monies, albeit voluntary contributions.

A special series of four state lotteries was introduced with profits earmarked for the proposed Opera House. Imposing a tax instead would have been highly unpopular and unlikely to win a clear majority amidst the somewhat grudging members of the State Caucus. Even the seemingly innocent lottery was not without its share of scandals. The first winner turned out to have been quite wealthy even prior to his win; another led to the first recorded Australian case of kidnapping for ransom. And this was even before any decision had been made to build the Opera House.

TROUBLES, TROUBLES, TROUBLES

Perhaps the first serious doubts about the practicality of Utzon's sketches were raised early in 1958 by the structural engineer of his choice. Let us consider their verdict directly (Yeomans, 1968: p. 51):

> *"The structural design of the (shells) is obviously quite a problem . . . it may even be necessary to connect the shells in some way to the internal structure of the halls . . . Extensive model tests will be required."*

The engineering firm, Arup, explained the reasons for this: With the winds blowing up the harbor, the "bending moments" could be quite considerable, an engineer's way of saying that with an improper design, the structure may blow away!

Utzon urgently needed a competent firm of structural engineers to help him build his creation. Within a day of his being declared the winner, he nominated Arup and Partners of London as the structural engineers with the blessings of the client, the Government of New South Wales. Utzon remained based in Denmark while working on the Opera House. He made visits to London and Sydney through the years 1958–1963, before taking up residence in Sydney amidst a royal reception. Three years later, his project still in the midst of construction, he resigned and left, with the SOH nowhere near completion! What happened in the interim? Unfortunately, a great deal. In an atmosphere of increasing rancour, including a series of three-way disagreements between the architect, the engineers, and the government, Utzon felt that he was being forced to shoulder the lion's share of the blame for delays while being ignored in decision making. Perhaps the straw that broke the camel's back was when the Department of Works delayed payment of Utzon's claims for stage techniques despite several month's reminders. The architect's resignation letter to the New South Wales government on February 28, 1966 speaks for itself (Yeomans, 1968: p. 148):

> *"As . . . you know . . . there has been no collaboration . . . in the last many months from your department's side. This also forces me to leave the job as I see clearly that you do not respect me as an architect."*

Utzon's resignation was accepted immediately "as a matter of regret," with the government at the same time paying tribute "to the whole concept" behind the architect's design. The Premier clarified that the fees in dispute dated back to 1960, though the formal claim for them was not made until 1965. If this is true, the incident which triggered the resignation letter really seems to be a nonincident, as the wording of the letter appears to bear out. Rather, the source of the disputes had been simmering for several years as both designer and engineering firm alike felt the pressure of lack of progress amid skyrocketing costs.

Despite all the problems, the SOH stands as a monument to its architect, even though he was not present at the 1973 inauguration.

Perhaps his own epitaph is best read in a brief quote from a knowledgeable source made over twenty years after the Opera House's completion (*Encyclopedia Brittanica,* 1993: p. 223c):

> *"[Utzon], best known for his dynamic, imaginative, but problematical design . . . placed first in the competition which brought him international fame . . . [As the project] posed a variety of problems . . . [he] resigned . . . in 1966 . . . [It] is now Sydney's best known landmark."*

COMPLICATIONS CONFOUNDED

The government sought to make an early start, lest the project be given up due to any major practical problems. To lessen the possibility of excessive scrutiny of the designs that could lead to shelving the project, the contract was initially divided into two parts and later into three distinct contracts, including those responsible for their oversight:

- Stage 1: The podium, everything up to start of roofs—Ove Arup & Partners
- Stage 2: The roofs—Ove Arup & Partners
- Stage 3: The rest, including interiors and machinery etc.—Utzon/Design team

Tenders for stage one construction attracted six Australian bids varying from $2.8–$4.5 million. Civil & Civic Contractors Pty. Ltd. of New South Wales, managed by a Dutch engineer, offered the lowest tender. They were awarded the job, and the physical work started on March 2, 1959. The early results were ominous and presaged the general problems that the government and Arup would have in staying within budget as the final cost for stage one construction ballooned to $5.2 million. How could a competent firm such as Arup allow this type of escalation? The results were partly due to the unsatisfactory nature of the contracts and coordination between Utzon and his design engineers, Arup. Utzon and Arup chose Danes as early consultants for stage techniques and electrical services. They were assigned their tasks by Utzon, but paid by Arup, to be reimbursed by the owner. Arup's engineers were unhappy about being forced to carry the legal responsibility and paying for work not personally controlled by them. This led to Arup's "Declaration of Independence" (March 26, 1963), which was quite revealing and illustrates the nature of the deteriorating relations between Utzon and his design engineers (Yeomans, 1968: p. 79):

> *"We, as structural consultants . . . [must] cast away pretense and make-believe and face the facts . . . of who is responsible for what . . . [O]nly [we] have a direct contract with the client . . . the specialist consultants (acoustics, electrics, heating etc.) are responsible . . . to the structural consultants . . . This makes it a kind of gamble whether the fees recoverable from the client are sufficient or not to cover the specialist fees . . . This situation is ABSURD (emphasis, ours) . . . [I]t is the architect . . . who chooses these specialists . . . [W]e are ourselves under the direction of the architect . . . So . . . we have responsibility without power . . . [serving as a] convenient victim, scapegoat, or stooge who can be blamed . . . but cannot possibly get any credit.*
>
> *[The] choice of Utzon's design was a controversial issue . . . whether it could be built at all . . . we have done far more for this job than could possibly be under our duties . . . [I]t has unavoidably fallen on us to work out many of the essentially architectural details . . . [All] this is having a detrimental effect on all our firm's activities."*

Arup's firm of consulting engineers specializing in buildings is the largest in Europe. As founder (1945), Arup himself serves as the main source of inspiration and is thought of as sort of a father figure. He opened new vistas in the use of concrete and created an atmo-sphere in which his staff worked, lived freely, and grew in their jobs. Known for leading without orders, anyone could argue with Arup and people were judged only by their actions. An engineer with imagination, Arup's relations with the client, as briefly described above, are one side of the story. An equally important side are his relations with the other major contractors, in particular Orup, the structural engineers.

Just about the time of their "Declaration," Utzon had been feeling that Arup was consistently not investigating his suggestions but instead offering unsolicited proposals, almost assuming the role of the principal architect. Naturally Utzon resented it. The tension between Utzon and Arup and his engineers kept mounting and the situation took a serious turn when they hit a snag with the acoustical ceiling. The government's oversight minister started dealing directly with Arup, which annoyed Utzon even further.

In a nutshell, the responsibilities of individual parties, in particular specialist firms, were not clearly spelled out. What is surprising, however, is that Arup pointed out this serious anomaly much too late, not before the completion of stage one. By this time, much of the damage had already been done, particularly with regard to the relationship between the architect and Arup. In retrospect, one is forced to ask the obvious question: How did Arup accept this position in the first place and keep silent for nearly four years? Arup's short and seemingly honest answer, "For the sake of the job," is not entirely satisfying. Certainly, there was a strong incentive to be involved in such a high-profile project. Further, the technical challenges of translating Utzon's "wonderful conception" and "dream" into a reality would appeal to any engineering firm. Nevertheless, the design limitations, coupled with serious flaws in project oversight, put Arup into an increasingly precarious situation, one from which he eventually rebelled.

TIME AND COST ESTIMATES

We have noted earlier that Utzon's entry included no design whatsoever, nor any indication how his "dream" could be realized in practice. The question of an estimate at that preliminary stage seems purely theoretical. But clearly, the estimate should have been a major deciding factor for the authorities in planning how they would finance the project. Consider some of the cost estimate numbers that were floated at various points in the development process, indicating the lack of systematic planning that the project received from its beginning:

Date	Estimate	Source of the Estimate
1957	$7 million	Joe Cahill's (Labour leader) first estimate
April '58	$9.6 million	Labour government's "official estimate"
April '61	$13 million	estimate from chairman of Technical Advisory Panel
July '62	$20 million	+ $10 million in items conveniently forgotten—Opposition Committee
Aug '62	$25 million	+ $12 million for additional items, including: $5.2 million "items not originally included," $2.2 million unexpected stage machinery costs
		$2.0 million more for roof than expected
		$2.4 million additional payment for stage 1 due to unexpected difficulties
		—Joint consultation between authorities, Utzon, Arup
May '64	$36 million	Opposition Leader
Aug '64	$49 million	Member of the Legislative Assembly
End '65	$100 million?	Minister of Works
Feb '67	$50 million	A government consultant

All these figures are taken from Yeoman's book which was published in 1968, about five years before the SOH was completed. We now know that the final price tag was $107 million. It seems that the Minister of Works, obviously well briefed by his advisors, had a realistic feel about the cost.

Clearly, the first figure of $7 million should not have been called an estimate, not even a valid guesstimate. Legitimate cost estimation can only arise when some basic design has been completed. In actual fact, at that time, no one had any idea whether the building as envisioned by Utzon could be built at all, and if so, exactly how. In such a case, the question of an estimate does not arise, certainly not to the degree that the judges would use it as criterion for selecting Utzon's original design.

IN RETROSPECT

The planned facility turned out to be much more than a mere opera house—it was designed as a major arts center, "beautiful and costly . . . a magnificent site" (*Encyclopedia Brittanica,* 1993: p. 360a). Sydney citizens obviously like the SOH to the extent that a team of public and private architects have since patterned the Overseas Passenger Terminal after it (Jahn, 1991).

However, all are not in agreement regarding the structure's aesthetic appeal and functionality. A seemingly knowledgeable expert has called the SOH the Mickey Mouse Opera House, thanks to the "lack of functional justification for the design," and used adjectives such as "not beautiful, nor ugly, . . . merely absurd . . . straight out of Disney World." (Waugh, 1988: p. 8) Even more damning is his observation (p. 8):

> *". . . the mystery of how a skeptical, satirically-minded nation allowed it to be built—at such prodigious cost, and with such flamboyant disregard for any canon of good taste or common sense . . . [indicates an] astounding level of tolerance."*

The Sidney Opera House debacle comprises a long story extending over fifteen years. There have been many reasons for the failure of the project. We have time and space here only to highlight some of the major factors rather concisely. Some of these may well pose more questions than provide answers to the "mystery" surrounding this (undoubtedly) controversial building!

- No hard-headed government should issue a start order for a project without the basic design and a realistic estimate of time and cost involved. Feasibility analysis was almost nonexistent as the judges made their selection on the basis of the drawing, without any follow-up analysis of the viability of the designs.

- No distinguished architect-judge should award first prize for a handful of sketches crammed with unsolved problems and without any basic designs.
- The resignation of an architect, in the midst of supervising one of his own designs for one of the great buildings of the century, suggests a seriously flawed development process, particularly regarding communication flows and accountability.
- No reputed structural engineers should work on a building before establishing that it is possible to construct the roof. The structural engineers were handed a fait accompli and expected to complete it to design specification. Obviously, this approach opened the floodgates to a number of structural and acoustical problems that had to be solved in crisis (reactive) mode. Engineers could not even develop adequate contingency plans because they were unable to accurately test out the architect's scheme for the building.

Although we have prefaced the above as statements, these are really vital questions which offer reasons for failure of the project. The result of this combination of factors? After nine years of planning and seven years of erection, the architect, his engineers, and also the client were at such loggerheads with each other that the work came to a standstill until the "horse" was changed midstream. And to add to this melodrama, all this happened in Sydney, where the goal of the Opera House project had been to make a cultural statement to the rest of the world, gainsaying the common misrepresentation of Australians as simply addicted to sun and frolic, while not caring for higher forms of art. Unfortunately, despite their attempts, the authorities concerned had no active support from the public. An appeal for funds to build the Opera House was a flop, but politicians were so sold on this idea that to perpetuate their names in history, they arranged to fund the project out of the profits of a state lottery. Ultimately the funding source did not matter: It could well have been funded by the state itself without making any difference to its final fate.

We will never fully know the reasons why Utzon resigned in the midst of the project, with his grand design beset on all sides by technical problems. Was it because he found it impossible to unscramble all the problems, some of which had become unbearably hard? Did the resignation actually give him relief? He seems to hint at answers to these questions in the wording of his resignation letter. In particular, the Minister's reply indicated that Utzon was pressing for payment of a claim which was formally submitted some four to five years AFTER the event, suggesting that it was more of a convenient excuse than a genuine trigger event. As to his competence in handling the job, we quote from one of Utzon's admirers, a Mr. Wheatland, who was himself a competitor but could not see how the specifications could be met (Yeomans, 1968: p. 57):

> *"Utzon is NOT an impractical visionary . . . [he] is very interested in structures and very knowledgeable . . . [He] can be a very good businessman, too . . . [His] general philosophy is that buildings should as far as possible be made of parts which had been constructed on the ground . . . [In fact] if Jorn had not been so interested in building techniques, there would have been no rows . . . at all . . . [H]e looked on prefabricated building as a way to obtain first-grade workmanship."*

However, to give a balanced picture, we should note the pronouncement of one of the Sydney area architects, Walter Bunning, in an article in the *Sydney Morning Herald* (quoted in Yeomans, 1968: p. 186):

> *"(SOH) started nine years ago, . . . unfortunately [under] a smokescreen. [Its] many failures are world-shattering . . . from a functional point of view (SOH) will be a second-rate building when compared with the Lincoln Center Opera House due to open in Oct. (1966) . . . seating more than SOH, with a final cost of $28 million, half the estimated cost for SOH . . . [It took] four years to build against ten if . . . it opens in 1969 . . . [A] comparison most unfavorable to SOH . . . [The] functional arrangements . . . [are] utterly unworkable . . . [as demonstrated by] the rise in cost from $7m to $50m . . . [including the] roof from original $1.6m to $10.6, [and the] rest of the building from $5.4m to $39.4m."*

References

Cargher, J. 1994. "Ports of Call," *Opera News*, vol. 59, November, pp. 24–28.

Encyclopedia Brittanica. 1993. vol. 28, Macropedia, p. 360a.

Encyclopedia Brittanica. 1993. vol. 12, Micropedia, p. 223c.

Jahn, G. 1991. "Terminal Truth," *Architectural Record,* vol. 179, pp. 204–211.

Kharbanda, O.P. and Stallworthy, E.A. 1984. *How to Learn from Project Disasters,* Aldershot, UK: Gower.

Waugh, A. 1988. "Some leisurely meditations on the Sydney Opera House," *The Spectator,* vol. 260, Feb 27, p. 8.

Yeomans, J. 1968. *The Other Taj Mahal—What Happened to the Sydney Opera House,* London: Longmans.

12

The European Aerospace Industry or A Study in Contrasts

FOUR MAJOR PROJECTS in the European Aerospace industry offer some constructive and unique lessons in project management, particularly the dangers posed by bureaucratic interference. The projects are Civil aircraft: Concorde and Airbus, and Space Organizations: ESRO and ELDO, the ancestors of the European Space Agency. We will compare and contrast these examples, drawing both a success story and a failure from each set of projects.

BACKGROUND: ECONOMIES OF SCALE

Since the 1940s, the European aerospace industry has undergone a change in respect to joint ventures, consortia, and partnerships—not all have been successful. For over two decades (the 1960s to the 1980s), the industry has tried hard to regain a defensible competitive position in the world markets through collaborative ventures, some highly successful, others true disasters. The two groups above offer an interesting study of contrasts in strategies, technology, and international relations, but most importantly (of particular interest here), they offer lessons in how NOT to manage projects—to avoid failure (Koening and Thietart, 1988).

The organizational forms used to run economic activities include a broad array, everything from arms-length transactions to the vertically integrated firm. These types constitute the extremes of the spectrum of forms. The "trick" lies in striking the right balance between the cost of organizing the systems and the costs of running them. Obviously, this balance shifts with changing circumstances. As an example, a firm in a nascent industry might reduce the risks by using an existing channel rather than creating its own distribution or supply channel. But for a novel or unique product, existing channels may not serve the purpose. Then there is the factor of economy of scale. It is usually more pragmatic economically to let one firm cater to several customers for a specific component, unless the firm is aiming at total vertical integration. Airbus (one of the four industries discussed here) had to set up its own sales force and keep it in close touch with its customers, since the existing aircraft manufacturers were already forwardly integrated.

Strategic choices for such major projects depend on three major factors:

- The nature of the competitive environment
- Their respective objectives
- Resources and capabilities at hand

The two successful projects, (Airbus and ESRO) had similar organizational features; so did Concord and ELDO, which proved to be commercial disasters. After laying down the necessary background, we will discuss these at length in pairs comprising a successful and a failed project in order to get to the root cause of their respective success and failure.

The aerospace industry has high asset specificity. It is also characterized by technical and economic risks. In the case of the aircraft industry, which is somewhat similar to the space industry, each major element, like the fuselage, cockpit, wings, and so forth is specifically designed and requires expensive and specific equipment, in addition to specially trained personnel. These are risky ventures indeed, marked with technical failures and prodigious commercial risks. The development period for a commercial aircraft averages five years, and for advanced design aircraft the development loop can be up to ten years. Likewise, payback ranges anywhere from ten to fifteen years, depending upon the initial success of the airplane. Clearly, by its very nature, predictions are hazardous.

There is another very relevant factor; namely, the economy of scale involved in aircraft production and sales. The break-even point of an aircraft program is generally considered to be 300–600 units. Clearly, no single European country affords a manufacturer such a large market; hence the need came about for a mutual organization to make such a project viable. The cooperation can take several forms as indicated here, and is further illustrated by the projects discussed later in this chapter.

For a commercial aircraft like the Airbus A-300, the typical development costs can be the sales equivalent of seventy planes—about $3.5 billion. For further product improvement, we need to add the sales equivalent of another forty planes, with a total risk of about $5 billion. Again, it is clear that no single European firm can afford to take such a huge risk. On the other hand, Boeing, the dominant airframe manufacturer, has historically been able to draw on resources from successful earlier generations, namely 727 and 737, to help finance the development of, say, the 747. Consequently, in order to be competitive in such a capital and technologically intensive industry, European companies have to pool risks in a mutual organization.

A popular form of cooperation is the main contractor/subcontractor type, with the major partner assuming overall responsibility for the design and production as well as for marketing and sales of the product. The subcontractors bear risk for their respective share of their work. A typical example of this in the commercial aviation field is Aeritalia of Italy. For some years now, Aeritalia has been a subcontractor for Boeing, particularly in the case of the jumbo jets, 747 and 767, supplying component parts under special licensing agreements. Unfortunately, this sort of relationship has increasingly become the exception rather than the rule with European companies, especially when a foreign partner (such as Boeing) joins with a local company. It was in order to ensure a steady market for aircraft component manufacturing that the mutual organization was originally formed. Did it prove to be the right choice? This is examined in the four case studies discussed here—all relating to the European aerospace industry.

CONCORDE: A SINGED BIRD

In the comparison of Concorde and Airbus, we have a study in contrasts between two major projects containing some valuable lessons in "dos" and "don'ts." Both the projects here relate to aircraft but with an important difference: Airbus is subsonic (below Mach 1, i.e., operating at a speed below that of sound, 660 miles per hour) whereas Concorde is supersonic, (i.e., flying at a speed above that of sound—above Mach 1). The latter is obviously fraught with more risks and uncertainties due to the technical issues to be solved. In retrospect, it seems that everything that could go wrong did go wrong.

The technical challenges relative to successfully developing a commercially-viable supersonic aircraft are daunting. Just to give you an idea of the severe conditions:

- At 70,000 ft, air is so thin, a person's blood will boil.
- At Mach 3 and -55C in outside air temperature, air friction on the aircraft's fuselage will generate a skin temperature of 345 degrees centigrade!
- At this speed, any instability could cause the aircraft to flip end over end and disintegrate in a matter of seconds.

- The unique nature of the physical environment in which the plane would fly necessitated that requisite materials of construction had to be developed specially.
- Fuel consumption is an astounding 8,500 gallons of a paraffin fuel mixture per hour.

In the late 1950s, the British Aircraft Corporation (BAC) and France's Sud Aviation started developing a commercial supersonic aircraft. Each favored a different type: BAC preferred an aircraft capable of long-range with Mach 2 performance. Sud opted for a medium-range (not cross-Atlantic) aircraft, a Mach 3 successor to the commercially promising Caravelle. Since they were dealing with vital projects of national importance (and prestige), the respective governments stepped in, and through a series of intense negotiations, helped bridge the gap between the two concepts. This resulted in an accord (reached in November 1962) for a seven-year joint project, based on an initial budgeted cost of $160 million. It was expected that 150–200 planes could be sold, so an initial modest production run of 100 was planned. The first test flight occurred in 1969, and the aircraft became operational in 1974, just at the time of the global "oil shock." The final cost to develop the airplane was 1.49 billion pounds, over fifteen times more than the first cost estimate of 95 million pounds (1959). Even after subtracting 350 million pounds in sales receipts, the total developmental expenditure was still a staggering 1.14 billion pounds.

In 1979, the Central Office of Information (COI) published consolidated data on Concorde that showed that the cost of the project, first estimated at 95 million pounds in 1959, had risen to some 700 million pounds by 1978, while sales had only brought in 350 million pounds. The early, wildly optimistic forecast that Concorde would be in service by 1970 was unrealistic from the start. The first production plane actually flew in 1973, while regular services were first offered in 1976. But that presented problems when the aircrafts were banned from American airspace as an environmental hazard—even though operating the cross-Atlantic route was precisely why the airplane had been developed! It was only after the president of France had personally intervened and made a direct approach to the president of the United States, that a limited number of flights were allowed to land in the U.S. for a test period.

Such enormous project overruns make a mockery of estimating. But let us be fair: Concorde was indeed a "development" project in every sense of the word. Its initial cost estimate resembled nothing so much as a "shot in the dark." Further, any estimator, or for that matter, any engineer or other project team member, is essentially helpless amidst the intrigue and hidden agendas surrounding political decisions. The project was ultimately halted in 1974 with a plan to build only sixteen planes, including four preproduction prototypes. Nine were sold, five to British Airways and four to Air France. On the whole, the aircraft represented a technological breakthrough, but a financial disaster. A post mortem revealed its reasons:

- Government indecision in regard to the two versions as well as the main project thrust. Right up until the project went into preproduction, there continued to be wide disagreement as to which of the two versions of the aircraft would be selected for development.
- The main driving force was to use the underutilized resources in the two neighboring countries, England and France. Concorde had many features of a "make work" program. Its primary goal appeared at various times to keep people employed and use government funds (derived from taxes) to support airframe component industries.
- Politics of being good Europeans overrode the techno-economic factors. The ultimate determinant used for resolving technical problems and disagreements was not to get the technology "right" so much as it was to maintain cordial relations. Nonhazardous decisions were made more on the basis of placating recalcitrant partners than keeping costs down and performance standards high.

- No real integrated management and a rotating chairmanship. Concorde was a project run by governmental bureaucrats. Ceremony and political expediency (convention) dictated strategic decisions with different foci coming from different chairmen.
- Political micromanagement at the very top. For example, President DeGaulle chose the colors for Concorde's seats; he also insisted on adding the "e" to the end of the airplane's name to give it a Gallic touch.
- A built-in mechanism for discord, delay, haggling, and duplicated costs. The project was based on the mutual organization mentality. While offering some obvious benefits, it also presents countries with frustrating features, since decisions are often made more for various national interests than with the intent of making the airplane as cost efficient as possible. When political expediency continues to outweigh pragmatic production decisions, the end result is waste.
- Cost-plus basis encouraged unnecessary design changes without any risk. Rather than use a fixed-cost contracting approach, cost-plus allowed for maximum flexibility and abuse of the system by subcontractors intent on wringing every bit of profit that they could get from the project.

CONCORDE IS DEAD. WHEN IS ITS FUNERAL?

The above was the terse comment heard in the newspapers after the over-ambitious production plans for Concorde evaporated in thin air. Such an evaluation is all the more ironic given the overly optimistic comment and almost universal praise for it's market potential during the development stage. There was, of course, no proper market survey. For that matter, no realistic cost estimate, let alone cost control was ever performed. Finally, the actual cost of development turned out to be nearly ten to fifteen times the original estimates.

Obviously, the prediction of design and development costs for civil airplanes clearly stated that estimating in this area was an inexact science (Harrold and Nicol, 1977). Further, Harrold and Nicol's research led them to conclude that even after thirty-five years of aircraft manufacture, too little information was available on past projects to be of much use in later projects.

As for the sales potential, the major airlines initially seemed enthusiastic at the prospect of owning and operating supersonic aircrafts. But the hefty price tag (presumably designed to cover the entire development costs) coupled with the timing of its introduction (coinciding with the worldwide "oil shock"), led to a change in this scenario. How expensive was the operation of the Concorde expected to be? The comment of a potential major airliner is truly indicative, stating: We can't afford the Concorde, even as a free gift!

SUPERSONIC MONSTER—A GLEAMING CINDERELLA

While Concorde continued to suffer through its development process, European politics mingled with and threatened to kill the project. Meanwhile, an almost simultaneous development was taking place across the Atlantic. The "supersonic monster" was the end product of the American R&D efforts towards their own supersonic aircraft. Described as "impossible and weirdly beautiful," it was also nicknamed (Wheeler, 1965):

- the flared triangle of wing
- six black caverns under the tail
- heron homesick for the sky
- cobra striking from on top of crane

In effect, this development envisaged a future-generations aircraft designed and built by the current generation. Test flights started at the end of 1964, with the usual quota of snags, false starts, and human errors. However, within a year of prototype development, and on its fourteenth test flight, it created a new record for continuous supersonic flight: eighty-four minutes, top speed 1,870 mph, altitude 68,000 ft. From a technical perspective, the engineers had systematically defeated the various challenges they faced in bringing such an aircraft to fruition.

Developed in America at a cost of approximately $400 million during the period from 1959–1967, the "monster" was designed to fly at 70,000 feet at a speed of about 2,000 miles per hour (Mach 3, or three times the speed of sound). First conceived in 1955 as a potential successor to the U.S. Strategic Air Command's B-52 heavy bomber, the project ended with the completion of a mere two experimental aircraft (the XB-70).

Just as the European Concorde was a technical triumph (though with far more flawed project development), the American SST ended up as a canceled project, retaining only the two prototypes that had been built. While there were a number of environmental protests centered on the airplane (its noise over residential areas, potential of damage from sonic booms, etc.), the major problem with the aircraft was its prohibitively expensive seat/mile expense ratio. There was simply no way to justify and maintain the aircraft given the fares that would have had to be charged to support it. Added to this was the impact of the oil embargo in the mid 1970s and the decision was clear: the design was shelved.

We cannot help comparing the simultaneous and independent American efforts with the British/French efforts in developing a somewhat similar supersonic aircraft. For about 1.5 billion pounds spent on the development and production of Concorde, the combined British/French team got fourteen operational aircraft. For one-third of that, about 500 million pounds, the Americans got nothing, in the sense that the project was dropped short of production. Obviously, there is a temptation to label one project as successful. In reality, however, it appears that both projects suffered from an uncertainty in terms of ultimate development. In other words, technological achievements aside, it is difficult to ascertain if there existed a national will to widely incorporate these aircraft into general production and use or if in fact they were intended to do little more than prove the viability of supersonic commercial travel (Henderson, 1977).

In retrospect, was such enormous effort and cost worth it? In the long run, hopefully, yes. The development process yielded a great deal of valuable knowledge, which may in due course lead to supersonic passenger travel. The agonizing years in design and construction for this great gleaming Cinderella plane may yet pay off in terms of a test model for commercial air transport.

LESSONS FROM FAILURE

There is no doubt that the Concorde design is a viable aircraft, and that flights at speeds of Mach 2 and over are of commercial interest, but the cost of its development and the uncertain strategic goals under which it was originally designed served to make it, in the final analysis, little more than a technically impressive oddity. The Concorde represented a huge waste of money and a profligate use of national resources. Yet in fairness, analysis of the project from a technical perspective shows it to be a generally successful example of a joint venture. Two countries, speaking different languages, with very different traditions and procedures, with a history of mutual antagonism, nevertheless successfully cooperated together (Powell, 1994). Their work pioneered international technical innovation and industrial collaboration on a grand scale, while restoring to Europe undisputed leadership in the most advanced field of commercial aircraft development.

There can be, and there probably are, a great many indirect benefits to society at large that will continue to flow from the new plateau of technology that was attained during the development of the Concorde. There was also a political aspect. The Concorde may well be seen as the price that Britain and France have paid to keep some of their most powerful and crucial defense contractors fully stretched. In effect, it is the price they paid for national security. The lesson for those involved in "development projects" is that the early estimates are always wrong; the magnitude of the error depends upon the degree of development necessary to achieve a successful result. And that is something no one knows before the event!

POSTSCRIPT: CONCORDE MAY YET FLY HIGH

We have taken you through the checkered history of the Concorde project from its inception to its development and introduction. Some of the headlines in various financial and aerospace journals in 1994, on the eve of the twenty-fifth anniversary of Concorde's maiden flight, held promise for the possibility of a brighter future for Concorde:

- "Growing Concorde charter service in North America"
- "Concorde cabin redesigned by the famous designer, Andre Putman"
- "NASA awards $880 million contract to Boeing and McDonnel Douglas to develop the next generation of supersonic transport aircraft"

Currently, British and French aviation officials are busily exploring methods for whether and how to extend the operating lives of their Concordes (Shifrin, 1994). This seems to suggest that despite having lost millions, reports of Concorde's demise are very much premature. In fact, a new generation of supersonic aircraft may be in the works. And standing on the shoulders of the pioneers in this field in Europe and the U.S., we may yet have a winner.

Also, having collaborated successfully on Concorde in the 1950s and 60s, Britain and France have in hand an intensive European technology effort to develop an improved version of a high speed transport aircraft (Sparaco, Proctor, and Smith, 1994). Hopefully, the lessons of the past (as described above) will be heeded in order to avoid similar pitfalls and mistakes. After all, this is our main message: If we take the lessons of "Cinderella" and Concorde to heart, the new incarnation of Concorde may prove to be a success. Or shall we go on repeating past history?

Phoenix Rising from the Ashes: Airbus

Standing in sharp contrast to Concorde's commercial failure is the singular success of Airbus. While it is not our purpose here to pursue an in-depth look at the Airbus story, its success is instructive in that it serves as an important counter-point to the problems experienced by Concorde. For example, the Airbus consortium consciously avoided many of the more obvious errors of their supersonic predecessor—apparently having learned some of the harsh lessons from the Concorde disasters. Although it actively received governmental subsidization during its development, Airbus and the respective European governments appear to have successfully decoupled political and nationalistic demands from the commercial development process. This allowed Airbus to make business decisions in an unfettered manner. The results have been resoundingly positive. Airbus is currently the second largest airframe manufacturer, behind Boeing. Having already captured 30 percent of the market for widebody aircraft, they were well placed for the new generation of single-aisle airplanes.

Perhaps the best summary of the differences between Concorde and Airbus was noted in an article that appeared at the time of Airbus' successful introduction (quoted in Koenig and Thietart, 1988):

> *"Concorde was an entirely political aeroplane: the plane was to show that we were good Europeans."*

> *"After Concorde, an engineer's dream built by politicians, comes Airbus, a businessman's dream built by engineers."*

ESRO vs. ELDO

The stories of ESRO and ELDO offer an interesting study in contrasts, as both agencies performed somewhat similar functions. ESRO's goal was to launch satellites, and ELDO was to develop a

European spacecraft. One succeeded, while the other failed. Why? Let us see what lessons they hold for project managers.

ESRO: The Virtual Mutual Organization

By the end of the 1950s, a need was felt to create a European Space Research Organization comparable to its counterpart in the nuclear field, CERN (Centre European de Recherche Nucleaire). ESRO was created in 1962 with a fully empowered director general backed by a council of two delegates (a scientist and a government officer) from each of the ten member countries.

Its initial objective was purely scientific: to run experiments with rockets and satellites over an eight-year period. Within broad guidelines, the director general had a great deal of freedom while operating under a three-year budget approved by the Council. ESRO was a legal entity and the director general had the power to appoint and dismiss top personnel, authorize acquisition of assets, and design and construct facilities. Home industries within each member country were enjoying a large share of their country's financial contributions, since 80 percent of it was required to be spent within each country. In other words, any country's contributions to ESRO practically benefitted subcontractors working within that country—a classic form of government subsidization of private business. This naturally encouraged industry's cooperation. Contracts were awarded to prime contractors who in turn awarded subcontracts to others.

In theory, this method of contribution and industrial subsidization seemed alluring, as it almost guaranteed minimal private resistance to ESRO. However, in actual practice, this procedure wasted a great deal of time and resources, and in due course created an enormous permanent bureaucracy within the ESRO organization. Contracting became difficult. Compounded with cost underestimation, there was a gradual shift of power within the organization and the Administration and Finance Committee assumed some of the powers of the director general.

Administrative costs escalated and the bureaucratic process remained supreme. In due course, by the mid 1960s, the once efficient ESRO organization found itself virtually paralyzed. A number of alternative solutions were proposed to remove these administrative roadblocks, to the point of outright abandonment of the agency in its current form. Fortunately, sounder wisdom prevailed and instead of turning to the various governments and their politicians for a bailout, ESRO turned within. A new director general was appointed, and with the help of a small advisory committee, he started to improve the financial situation, concentrating on streamlining the contracting process and removing excessive administrative hurdles. This step made all the difference. The mutually beneficial ESRO-industry relationships were strengthened. With the necessary expertise and decision-making power, ESRO was able to deal effectively with a set of competent prime contractors. In essence, the new director general successfully shifted the organization's focus back to its main intended thrust—the development of applied space research rather than a creeping bureaucracy and "staff infection." Once the process was back on track, and the central goals of the organization were reinforced, scientific success was achieved, although at times, cost exceeded the initial estimates. They were no doubt working in uncharted waters.

ELDO: The Stillborn Mutual Organization

Ultimately a study in contrasts, the success of the ESRO concept gave rise to the creation of ELDO. ELDO was responsible for the development and administration of a European spacecraft. Promoted initially by the British and the French, others joined in, including Germans, Italians, Belgians, and Dutch, each sharing part of the total responsibility. Although their structures and mandates were similar, marked differences between ELDO and ESRO played a significant role in its ultimate demise:

- Rather than developing a central oversight body, supervision of various components of the agency remained at the country level. ELDO rejected the idea of central control, allowing each country to operate its own processes, often with disastrous results as national expediency routinely overcame practical project requirements.

- Contracts could only be placed in a country with the approval of the country concerned. This procedural decision led to inefficiencies, as governments often awarded contracts to higher bids.
- Because of the allure of spacecraft at the time, there were far more politicians and bureaucrats in ELDO, unlike the scientists who principally populated ESRO. ELDO suffered from its own high visibility, attracting politicians in droves who all insisted on adding their input to decision processes, uninformed though it often was.
- ELDO was headed by a professional bureaucrat, unlike ESRO, which was run by a scientist. ESRO was able to defeat the creeping tide of bureaucracy because it maintained a goal-oriented scientist as its leader. ELDO, run by a professional bureaucrat, soon became a breeding ground for additional bureaucracies, adding departments and oversight committees seemingly by the dozen, resulting in a stranglehold on response time and innovative options.

The problems became only more compounded at the strategy-making level. In decision-making forums, ministry officials dominated, unlike ESRO, where scientists predominated. The demands and requirements of the national industrial firms were conveyed to ELDO by the respective governments without any real evaluation. Further, the ELDO structure became an anomalous assemblage of national programs where various government priorities remained preeminent. This was further confounded by the adoption of a rule of unanimity on substantive issues; that is, any material questions had to be approved by a unanimous vote of delegates from each country, almost ensuring that the process would remain hamstrung.

What was the final result of the creation of such bureaucratic hurdles and administrative interference? The budget was successively revised from $197 million to $300 million, with later figures escalating to $425 million, $626 million, and finally to $720 million. At that point, France challenged the viability of the project, saying that the rockets in development already seemed obsolete. In disgust, Britain withdrew its own support, leaving other member nations to carry on despite numerous problems and handicaps.

In due course, ELDO led to ESA, the European Space Agency. While indications suggest that its current form will be more successful, it is clear that the first attempt constituted a financial and performance failure for a number of reasons, including:

- No clear direction from the top. Politics and self-interest ruled over scientific process. Decisions could not be made quickly due to the number of bureaucratic hurdles set in place. Further, these decisions were often in the form of political compromise, rather than scientific objectivity, which aimed at seeking the best answer.
- Parochial goals prevailed over the organization's goals. Each country set its concerns higher than the concerns of the community.
- Decisions were rapidly passed up the ladder to ministerial levels, where they usually sat, were debated, and sat longer. Bureaucratic paralysis was the order of the day—day after day!
- Failure to build an operational and multinational organization. ELDO was never a consortium in the classic sense; rather, it was an amalgam of piece work done by organizations in different countries, all under their own government's control, with little attempt made to blend and develop synergies through other organizations.
- Crisis and escalating costs led to low commitment to the final product as various countries questioned the value of their own participation and began dropping out or scaling back their support. When problems occurred, each country's

representatives seemed more intent on establishing a protective layer between themselves and the problem. Rather than working to fix the problem, the mood was more one of seeking to fix (or avoid) the blame.

LESSONS FOR MANAGERS, ENGINEERS, AND GOVERNMENTS

Having discussed the four major European projects in the field of aerospace, we move now to the crucial part of the lessons. Of the projects discussed in this chapter, Concorde and ELDO were failures (though Concorde was a technical triumph), and Airbus and ESRO proved highly successful. So what are the lessons?

There are lessons here for all those who were involved in the projects, namely, managers, engineers, and government officials. Several factors are responsible for the success experienced by Airbus and ESRO. The following points have been highlighted:

Put professionals in charge and allow them to operate as the real champions.

It sometimes seems that one of the unfortunate axioms of technological development is that the more involved professional bureaucracies become, the smaller the likelihood of creating anything of significance. While perhaps an overly cynical view (although one fueled with much practical experience!), the underlying freedom of action and capacity for rapid response that are so fundamental to successful innovation are too often stifled by the excessive control mechanisms spawned by political oversight. Certainly programs of national importance are likely to attract their share of political interest and attempts at interference, but as shown in the case of ELDO, the reconfiguration of this agency into one run by bureaucrats provided the exact wrong messages.

It is ironic to consider that one of the oft-touted benefits of professional administration lies in its ability to keep a close lid on costs. That is, as such administrators display the natural fear within many engineering-dominated firms, cost considerations are often lost in pursuit of technology for technology's sake. The case of ELDO presents a refreshing reanalysis of this core belief. Right or wrong, there is no denying that much of the cost spiral experienced by ELDO came about not through excessive technical dabbling but through governmental interference.

Organizations must be flexible and adaptable to respond to their environment.

Coupled with the experiences discussed above is the obvious addition: when organizations are granted the flexibility and relief from too-tight control, they can perform in a responsive manner. It is often only in the cases where such control is overly formal that it also becomes overly restrictive and ultimately, counter-productive.

Relative decentralization, with short and rational decision-making processes, is a strong impetus for innovation.

Organizations that are routinely subjected to intense micro-management, particularly from politicians feeling the pressures common to nationalistic vagaries, find it incredibly difficult to operate in innovative and risk-taking modes. On the other hand, when professionals are given a clear mandate with little interference in day-to-day affairs, there is a much greater likelihood of positive results.

Satisfactory coordination and integration between the partners is vital.

Another major finding from these cases studies was the importance of establishing clear lines of communication among the various parties involved, particularly when those parties represent the interests of various national governments. Unless some conflict resolution mechanism is adopted a priori to project development, it is likely that excessive time will be spent, not on project-related activities, but in resolving the disagreements an goal priorities of member groups.

Roles must be well-defined with a shared view of organizational goals.

Project disasters can often be traced to a goal "vacuum." The more ambivalent or broadly-defined a set of goals, the greater the difficulty in achieving anything of importance. Ambiguity creates an atmosphere in which simple control mechanisms cannot operate. In the successful case studies, the goals driving both the projects themselves and the overall organizational mandates were specific enough to serve as a rallying point for the disparate groups and nationalities. On the other hand, in the absence of a clear and ordered set of goals, the development process became of greater and greater importance while the actual purpose shrunk in importance. To paraphrase Marshall McLuhan, it was the project development equivalent of the dictum: the medium is the message.

Conflicts should be resolved quickly within the organizational structure.

Another characteristic of successful development was the implicit belief that project-related difficulties were just that—project-related. When problems occurred, they did not serve as the basis for international squabbling or as a starting point for politicians and bureaucrats to make ad hominem attacks. Rather, conflict was recognized as a side effect of high technology project development and was dealt with in a professional manner, *within* the confines of the project's organization.

In the ultimate analysis, the paramount factors were the end results and the "bottom line."

Successful projects are usually driven by a clear desire to attain the ultimate goals. These successes were no different. The budgetary and performance "bottom lines" served as the superordinate goals that in effect were the impetus to resolve any side bar disagreements. In other words, for both Airbus and ESRO, the organizations adopted a clear "eyes on the prize" mentality that allowed them to adjudicate and resolve any technical, human resource, budgetary, or nationalistic issues in a satisfactory manner, without having to sacrifice the basic goals of the projects.

The above set of factors were largely absent in the case of the failed projects, Concorde and ELDO. Politicians were in charge of those projects instead of engineers, and they kept busy politicking. In a nutshell: adaptation is the secret to success, in sharp contrast to maintaining stability. While the latter is a short-term response and often the easy way out, perpetuating the status quo with regard to innovative projects is invariably the wrong route. The right decisions are usually the most difficult to make. Just as in decision-making, national will and personal courage are required in cases where payoffs are measured in terms of decades, rather than in time for the latest by-elections. When this truism is remembered, we can look forward to technological progress. But when it is forgotten, we continue to seed a frustrating and ultimately wasteful cycle.

References

Harrold, K.G. and S.I.R. Nicol. 1977. "The prediction of design and development cost of civil airlines," a paper presented at the Royal Aeronautical Society Management Studies Group Symposium on "Cost Prediction," March.

Henderson, P.D. 1977. Two British errors: their probable size and some possible lessons," *Oxford Economic Papers,* vol. 29, pp. 159–205.

Koenig, C. and R.A. Thietart. 1988. Managers, engineers, and government: The emergence of the mutual organization in the European Aerospace Industry, *Technology in Society,* 10, 45–69.

Powell, C. 1994. "A fundamental incompatibility," *Spectator,* vol. 273, Sep 3, pp. 8–10.

Shifrin, C.A. 1994. "Concorde facing life-extension verdict," *Aviation Week & Space Technology,* vol. 141, Nov 21, pp. 60–61.

Sparaco, P., P. Proctor, and B.A. Smith. 1994. "France, U.K. promote European HSCT effort," *Aviation Week & Space Technology,* vol. 141, Nov 21, pp. 62–69.

Wheeler, K. 1965. "The Birth of a Supersonic Monster," *Readers Digest,* Nov, pp. 128+, 5p.

13

The Tacoma Narrows Suspension Bridge

INTRODUCTION

The Tacoma Narrows suspension bridge (the third largest in the world after the Golden Gate and George Washington bridges) collapsed over fifty years ago through wind-induced failure, just four months after opening to traffic. When originally conceived and constructed, the Tacoma Narrows Bridge represented a triumph of modern engineering, using the most up-to-date materials and scientific construction techniques. Nevertheless, the bridge was plagued by serious design flaws which, given the special weather conditions of the Northwest, led to almost immediate catastrophic failure. A movie taken by chance during the disaster has contributed significantly to our understanding of the causes and ultimately led to a major reassessment of the manner in which engineering bridge design is undertaken.

BUILDING BRIDGES: THE CHALLENGES CONTINUE

With over one million bridges in regular use in the U.S., perhaps one out of every five is vulnerable to failure. To reconstruct all of them may cost billions. The percentage of bridges at risk may be slightly lower in Europe and Japan due to (hopefully) conservative design and higher safety factors.

Catastrophic bridge failure is by no means isolated or uncommon. However, it is important to distinguish between bridge failures during construction and those that fail at later points as a result of overuse, stress, and poor maintenance. While the later case may also cause tremendous damage, it is not usually tied to any identifiable project failure linked with the bridge's original construction. The first case, involving bridge collapse while under or immediately proceeding completion is a much more jolting reminder of the seriousness with which bridge construction must be approached, particularly in cases where the bridge design represents an innovative departure from traditional design or building methods.

One classic example of project-related failure in bridge construction is the Great Bridge in Melbourne. The bridge was intended to be part of a modern eight-lane highway linking the city's east and west sides. For all appearances, the bridge appeared strong and well-constructed, but deep within, it had a fatal sickness and became a crumpled ruin even before it was completed. Work on the project began in April 1968, using British design, Dutch steelwork, and Australian concrete, by companies that were leaders in their respective fields. However, due to strikes and poor on-site supervision, the project was running seven months behind schedule by the end of 1969. The underlying atmosphere at the work site was full of great antagonism between the unionized workforce and management groups. Imagined slights often brought work to a halt. For example, when one day's lunch servings consisted of fish and chips instead of the ordered hamburger, the union members organized an immediate work stoppage in protest!

Meanwhile in June 1970, a bridge of somewhat similar design being built in Wales in the U.K. (Milford Haven) collapsed while under construction, causing the deaths of four workers. Even this forewarning was ignored or minimized by the Melbourne construction crews. Although the first signs of failure became gradually apparent in early September, work continued on the span. The following month (October 15), the bridge suddenly crumpled, killing thirty-five of the sixty-eight men on the worksite at the time. The inquiry commission's verdict was damning in its apportionment of blame for the disaster, stating that to a greater or lesser degree, the authority itself, the designers, the contractors, and even the labor groups engaged in the work, must all take some part of the blame. In the final analysis, however, the greater part of the blame must be attributed to the designers who created a span with fatal structural weaknesses.

The bridge's original designers were dismissed and lessons were hopefully learned by all parties. A new design firm was hired, and using a more conventional design, the Melbourne bridge was redesigned and completed in 1974. The failure sent danger signals all over the world and as a result, in the U.K. alone, sixteen of the fifty-one bridges using a structural design similar to the original Melbourne model were ordered to be strengthened, and another sixteen were placed under closer scrutiny.

Bridge builders do indeed learn more from their failures than from their successes. The only redeeming feature in the case of the Great Bridge was that it fell prior to and not after completion and commissioning, with large numbers of commuters moving on its carriageways. During the same two year period, November 1969 to November 1971, at least three other box-girder bridges collapsed:

- Fourth Danube Bridge, Vienna
- West Gate Bridge
- Koblenz Bridge, West Germany

As with the Great Bridge collapse of Melbourne, comprehensive studies were conducted following each of these bridge disasters. A common factor found among the four failures demonstrated quite clearly that safety margins were not large enough. Engineers had willfully cut corners to lower the cost and had therefore taken an unknown and uncalculated risk.

POEMS IN STEEL AND CONCRETE

Bridges have quite appropriately been called "poems in steel and concrete." (Fleming, 1978) They look strong and are beautiful symbols of our advanced civilization, but as pointed out above, the beauty may be hiding disaster. These problems may be due to a design fault, skimping on the safety factor, or using inferior materials in order to reduce cost. Alternatively, (as is often the case) corrosion and general wear and tear take their toll. Consider a few typical examples:

On a foggy, rainy night in February 1975, a car skidded and hit a steel support beam of the Yadkin River Bridge near Siloam, North Carolina. The steel structure, weakened by years of use and inadequate maintenance, quickly gave way, hurtling the 225-foot span into the river, and with it, the occupants of six cars plunged into the water. The final toll was four dead, sixteen injured.

In May 1973, an elevated section of Chicago's Dan Ryan Expressway suddenly sagged thirteen inches. Fortunately, this happened early in the morning, well before the rush hour. Otherwise, with the normal load, the four-lane bridge structure would have collapsed, sending scores of cars and trucks crashing some fifty feet to the ground.

The Silver Bridge over the Ohio River (between Point Pleasant, West Virginia and Kanaungo, Ohio) collapsed. In one of the worst bridge disasters of the 1960s, thirty-one cars plunged into the river, leaving forty-six dead.

Of the over one million bridges in the U.S., on the average a bridge sags, buckles, or collapses every other day. Most of these are modest structures spanning across countryside creeks, but some are major structures with high stakes. Pittsburgh, for example, is popularly known as the "city of bridges," thanks to its site along three rivers with numerous spans. Nearly a fourth of the bridges across the major rivers, plus another 100 in the area have been labeled deficient. Closing bridges because of structural unsoundness leads to expensive detours for trucks and buses, traffic jams, and wasted time and resources. But the alternative, operating bridges past their safety capacity, is an invitation for tragedy. The situation in other large American cities, including New York and St. Louis, is not considered to be any better. The state of bridges in rural areas seems to be equally, if not more serious. In some of the counties, load hazards force children to exit their school buses and walk across a bridge, while their empty bus follows them.

Why do bridges deteriorate? Age is naturally a major factor, followed closely by sheer neglect. Even the best built bridges suffer from the effects of metal fatigue over long periods of time, requiring a planned program of upgrading and maintenance. When water seeps through cracks, steel bars rust, and as water freezes and thaws, concrete breaks into pieces. Potholes develop, chunks break off, and the reinforcement bars are exposed to the elements. Salt used to cope with heavy snow, acts as a poison, equivalent to the effects of battery acid. In the process, the steel corrodes. Then there is the usual overload, invariably well above the design capacity. For example, Chicago's Dan Ryan Expressway was designed for 150,000 cars a day, but at the time of the sagging, it was probably carrying a 50 percent overload.

Nevertheless, activity in building bridges continues unabated. After all, bridges are indispensable, and as yet there is no substitute for them. The "poems" continue to be made bigger and better. One of the most ambitious projects in the world is currently being planned, and it is a truly immense undertaking. The plan is to build a bridge to connect Shikoku, the smallest of Japan's four main islands, with Honshu, the largest. To accomplish this feat will require all the latest and best bridge-building techniques. For example, the current plan calls for eighteen major bridges to make up the three routes between the islands. The final structure is designed to be the longest cable-stayed crossing in the world with a center span 2,920 feet in length, eclipsing the current world's largest: France's 2,808 foot Normandie Bridge (Normile, 1994). Another equally impressive project that is in its early development stage is the proposed bridge link between Nova Scotia and Prince Edward Island in Canada, a structure that would span miles across the unpredictable Bay of Fundy.

THE TACOMA NARROWS SUSPENSION BRIDGE: "SOUND STRUCTURE!"

The dramatic collapse of the Tacoma Narrows suspension bridge in 1940, barely four months after completion, was a severe blow to the design and construction of large span bridges. It serves as a landmark failure in engineering history and is, indeed, a featured lesson in most civil engineering programs. The story of the collapse is well-known and in retrospect, serves as a fascinating account of one important aspect of project failure—engineering's misunderstanding of the effect that a variety of natural forces can have on projects, particularly in the construction industry.

Opening in July 1941, the Tacoma Narrows Bridge was built at a cost of $6.4 million and was largely funded by the Federal Government's Public Works Administration. The purpose of the bridge was essentially viewed as a defense measure to connect Seattle and Tacoma with the Puget Sound Navy Yard at Bremerton (*New York Times,* 1940). As the third largest single suspension bridge in the world, it had a center span of 2,800 feet and 1,000 foot approaches at each end.

Even before its inauguration and opening, the bridge began exhibiting strange characteristics that were immediately noticeable. For example, the slightest wind could cause the bridge to develop a pronounced longitudinal roll. The bridge would quite literally begin to lift at one end and in a wave action, the lift would "roll" the length of the bridge. Depending upon the severity of the wind, cameras were able to detect from anywhere up to eight separate vertical nodes in its rolling action. Many motorists crossing the bridge complained of acute seasickness brought on by the bridge's rising and falling! The strange weaving motion of the bridge became so well known to the locals that they nicknamed it "Galloping Gertie."

It was clear to all involved in the project that the bridge was experiencing increasing and unexpected difficulties. In fact, the weaving motion of "Galloping Gertie" became so bad as the summer moved into fall that heavy steel cables were installed externally to the span, in an attempt to reduce the wind-induced motion. The first attempt resulted in cables that snapped as they were being put into place. The second attempt, later in the fall, seemed to calm the swaying and oscillating motion of the bridge initially. Unfortunately, they proved incapable of forestalling the effects that the dynamic forces (wind) played on the bridge and they snapped just before the final critical torsional oscillations that led to the bridge's collapse.

On November 7, 1940, barely four months after opening, with winds of forty-two miles per hour blowing steadily, the 280 foot main span, which had already begun exhibiting a marked flex, went into a series of violent, vertical and torsional oscillations. Alarmingly, the amplitudes steadily increased, suspensions came loose, the support structures buckled, and the span began to break up. In effect, the bridge had seemed to come alive, struggling like a bound animal, and was literally shaking itself apart. Motorists caught on the bridge had to abandon their cars and literally crawl off the bridge as the side-to-side roll had become so pronounced (by now, the roll had reached forty-five degrees in either direction, causing the sides of the bridge to rise and fall over thirty feet) that it was impossible to traverse the bridge on foot.

After a fairly short period of time in which the wave oscillations became incredibly violent, the suspension bridge simply could not resist the pounding and broke apart. Observers stood in shock on either side of the bridge and watched as large pieces of the roadway followed by entire lengths of the span rained down into the Tacoma Narrows below. Fortunately, no human lives were lost, since traffic had been closed in the nick of time (one motorist accidentally forgot the family pet, a Cocker Spaniel, in his car, which went down with the bridge).

The slender twelve meters wide main deck had been supported by massive 130 meters high steel towers comprised of 335 foot long spans. These spans managed to remain intact despite the collapse of the main span. The second bridge (TNB II) would end up making use of these spans when it was rebuilt shortly thereafter, with a new span stiffened with a web truss.

Following the catastrophic failure, a three-person committee of top scientists was immediately convened by John Carmody, administrator for the Federal Works Agency, to determine the causes of the Tacoma Narrows collapse. The board consisted of some of the top scientists and engineers in the world at that time: Othmar Ammann, Theodore von Karman, and Glenn Woodruff. While satisfied that the basic design was sound and the suspension bridge had been constructed competently, they nevertheless were able to quickly uncover the underlying contributing causes to the bridge collapse.

First, the physical construction of the bridge contributed directly to its failure and was a source of continual concern from the time of its completion. Unlike other suspension bridges, one distinguishing feature of the Tacoma Narrows bridge was its small width to length ratio—smaller than any other suspension bridge of its type in the world. That ratio means quite simply that the bridge was incredibly narrow for its long length, a fact that was to contribute hugely to its distinctive oscillating behavior. Although almost one mile in length, the bridge was only constructed to carry a single traffic lane in each direction.

Another feature of the construction that was to play an important role in its collapse was the substitution of key structural components. The chief engineer in charge of construction, Mr. Charles Andrews, noted that the original plans called for the use of open girders in the construction of the bridge's sides. Unfortunately, at some point, a local construction engineer substituted flat, solid girders which deflected the wind rather than allowing for its passage. The result, Mr. Andrews noted, caused the bridge to catch the wind "like a kite" (*New York Times*, 1940: p. 3) and adopt a permanent sway. In engineering terms, the flat sides simply would not allow wind to pass through the sides of the bridge, reducing its wind drag. Instead, the solid, flat sides caught the wind, which pushed the bridge sideways until it had swayed enough to "spill" the wind from the vertical plane, much as a sailboat catches and spills wind in its sails.

A final problem with the initial plan lay in the actual location selected for the bridge's construction. While the investigating committee did not view the physical location of the bridge as contributing to its collapse, it did play an important secondary role through its effect on wind currents (United States Federal Works Agency, 1941). The topography of the Tacoma Narrows, over which the bridge was constructed, was particularly prone to high winds due to the narrowing down of the land on either side of the river. As a local engineer suggested, the unique characteristics of the land on which the bridge was built virtually double the wind velocity and act as a sort of wind tunnel (*New York Times*, 1940).

Before this collapse, not much was known about the effects of dynamic loads on structures (von Karman, 1963). Until then, it had always been taken for granted in bridge building that static load and the sheer bulk and mass of large trussed steel structures were enough to protect them against possible wind effects. It took this disaster to firmly establish in the minds of design engineers that dynamic, and *not* static loads are really the critical factor in designing such structures.

The bridge had been designed and built by some of the leading bridge engineers of America. They were well aware of the implications of the physical conditions where the bridge was to be constructed. The Tacoma Narrows area is known for strong and prolonged winds. Even so, the engineers were unworried: The oscillation characteristics of bridges under high winds were well-known and had been calculated into the design configurations. However, these oscillations in the Tacoma area, as measured by local engineers, were found to have amplitudes of as much as 1.5 meters. Despite such serious oscillations, the bridge's New York designer, Leon Moisseiff is said to have proclaimed with confidence that the structure was sound. It is no wonder then, that its collapse, as seen in the detailed pictures and film, came as a severe shock to the civil engineering community worldwide. In fact, the morning after the collapse, Moisseiff was quoted as saying, "I'm completely at a loss to explain the collapse" (*New York Times,* 1940: p. 3)

The engineering profession took these lessons to heart and set about a radical rethinking of their conventional design practices. The stunning part of this failure was not so much the oscillations, but the spectacular way in which the wave motions along the main span turned into a destructive tossing and turning and led finally to the climax in which the deck was wrenched out of position. The support cables snapped one at a time and the bridge began to shed its pieces in larger and larger chunks until the integrity was completely compromised.

Purely by coincidence, Professor Farquharson of Washington University happened to be on the bridge on the fateful day. He had been conducting extended experiments on a model of the bridge in order to possibly control its well-known oscillations. As part of this investigation, he came to the bridge to record the abnormal movements which had generally been observed and duly reported. While recording these, the professor suddenly observed the abnormal movements and realized that the bridge's oscillations were speeding up and becoming more pronounced. As the torque level steadily increased, Farquharson understood that the worst could, and actually *was,* happening. He immediately rushed to a nearby camera shop, borrowed a movie camera and began filming the bridge from different positions and angles. This factual and highly graphic record of the collapse as it happened, proved most useful for later investigations, including the post mortem of the disaster.

TACOMA NARROWS BRIDGE: THE POST MORTEM

Immediately following the bridge's collapse, local engineers had some strong hunches as to the cause. The chief engineer, Mr. Andrews, was quoted by the *New York Times* (1940: p. 3) as saying:

> *"In my opinion—and it's purely opinion—it (the collapse of Tacoma) was due to the web-girdered stiffening trusses. These caused the bridge*

> *to flutter, more or less as a leaf does, in the wind. That set up a vibration that built up until the failure occurred."*

Promptly thereafter, the Federal authorities set up their own board of engineers to investigate the cause of the failure, as noted above. After seeing the film of the bridge's collapse, von Karman wrote to strongly caution the authorities in light of the designer Moisseiff's comment (before the collapse) that the bridge structure was sound. von Karman was particularly incensed at Moisseiff's declared intention (after the collapse) to rebuild the bridge to the original design!

The board's final report placed the blame squarely on the inadequacy of a design that did not anticipate the dynamic properties of the wind. While longitudinal oscillations were well understood and had been experienced early in the bridge's construction, it was not until the bridge experienced added torsional rolling movements that the failure became inevitable (University of Washington Structural Research Laboratory, 1949).

The board concluded that the initial failure was due to (United States Federal Works Agency, 1941: p. 1):

> *"the slipping of the cable band . . . to which the center ties were connected. This slipping probably initiated the torsional oscillations . . . (which) caused breaking stresses . . . and further structural damage followed. The dropping of the . . . span was [due] to failure of the suspenders . . . followed by the sudden sagging . . . [Overall] it was not realized that the aerodynamic forces which had proven disastrous in the past to much lighter and shorter flexible suspension bridges would affect a structure of such magnitude."*

Bridge engineers, trained to think only in terms of static loads when designing bridges, found it hard to accept the crucial nature of these newly discovered dynamic forces. Von Karman, a leading expert on aerodynamics, butted heads more than once with other members of the board investigating the collapse. At one point he was asked point-blank to estimate the load forces acting on the bridge. When he gave his figures, the other engineer exclaimed in triumph (von Karman, 1963: p. 214), "See, we have calculated our design for a load three times that much." von Karman retorted that the engineer was still thinking in terms of static loads, whereas he had been calculating aerodynamic forces. Even small loads can be extremely dangerous when causing structural oscillations. On another occasion, his colleagues contemptuously dismissed von Karman's suggestion that wind-tunnel testing be used on new bridge design. von Karman clearly understood the principles of structural aerodynamics, but found it extremely difficult to convince his fellow engineers on the board. It is in this context that he later wrote his memoirs in which he proclaimed his dilemma in this regard (von Karman, 1963: p. 214):

> *"Bridge engineers, excellent though they were, couldn't see how a science applied to a small unstable thing like an airplane wing could also be applied to a huge, solid, nonflying structure like a bridge."*

Use of von Karman's analysis helped explain the collapse of suspension bridges due to high winds. Based on this, as expounded in his book noted above, the study of airflow and wind velocity helped make possible the design of aerodynamically stable bridges. It is perhaps ironic that it took the catastrophic failure of the bridge to advance the science of bridge building and create an entire new field of engineering through the study of dynamic (particularly aerodynamic) forces on structures. Again, this serves to reinforce the point of this book, that it is precisely through our failures that mankind is often disposed to extend the boundaries of human knowledge, learning from these mistakes and modifying behavior accordingly.

The lessons from the Tacoma Narrows bridge collapse are primarily those of ensuring a general awareness of technical limitations in project design. As Petroski (1985) has noted, the advances in technology often lead to a concomittant willingness to continually push out the edges of design envelopes and to try and achieve maximum efficiency in terms of design. The problem with radical designs or well-known designs used in unfamiliar ways is that their effect cannot be predicted using familiar formulae. In essence, a willingness to experiment requires that designers and engineers work to simultaneously develop a new calculus for testing these designs. It is imperative to understand the close relationship between new approaches and the need for equally new approaches to benchmarking and testing. It is dangerous to assume technology that works well in one setting will work equally well in another, particularly when other variables in the equation are subject to change.

The Tacoma Narrows Bridge collapse began in high drama and ended in farce. Following the bridge's destruction, the state of Washington attempted to collect the $6 million insurance refund on the bridge. At that time, it was discovered that the insurance agent had simply pocketed the state's premium and never bothered to obtain a policy. After all, who ever heard of a bridge the size of the Tacoma Narrows span collapsing? As von Karman wryly noted (1963: p. 215), "He (the insurance agent) ended up in jail, one of the unluckiest men in the world."

References

Fleming, T. 1978. "Our Bridges Are Falling Down, *Reader's Digest,* Nov., p. 209+, 5p.

New York Times. 1940. "Big Tacoma Bridge Crashes 190 Feet Into Puget Sound," Nov. 8, pp. 1–3.

Normile, D. 1994. "A short course in modern bridges," *Technology Review,* vol. 97, Nov–Dec 1994 pp. 52–9.

Paul, A.M. 1972. "The Day the Great Bridge Fell," *Reader's Digest,* Aug, pp. 86-91.

Petroski, H. 1985. *To Engineer is Human—The Role of Failure in Successful Design.* New York: St Martins Press.

United States Federal Works Agency. 1941. *The Failure of the Tacoma Narrows Bridge.* Pasadena, CA: Board of Engineers.

University of Washington Structural Research Laboratory. 1949. *Aerodynamic Stability of Suspension Bridges With Special References to the Tacoma Narrows Bridge.* Seattle, WA: University of Washington Press.

Von Karman, T. 1963. *The Wind and Beyond.* Boston, MA: Little, Brown.

14

The Nuclear Graveyard

INTRODUCTION

Thanks to safety and environmental concerns, a number of completed nuclear power plants have had to be abandoned, notably one each in Austria and the Philippines, and several in the U.S. The enormous time and cost overruns associated with their construction have also contributed to this result, especially in the U.S., where some of the plants are being converted for conventional fuel use. The Austrian plant was abandoned because a referendum on the eve of its commissioning narrowly voted against the start up. The Philippine plant has been mothballed amid charges of corruption and illegal payoffs by the plant's main contractor, Westinghouse, to the former president, Ferdinand Marcos. The story in the U.S. is long enough to warrant a full length book, but we bring you just a bit of the "flavor," unpalatable though it may be.

Not all nuclear plant construction is doomed at its inception.

France and Japan, in particular, have been very successful in building plants efficiently, so that the end result is a viable installation, well able to compete in terms of cost of electricity with the alternative sources. But management must be in control. An incredible story is told of an incident at TVA's Brown's Ferry plant in Alabama, where the reactor was seriously damaged by a fire caused by a technician testing equipment for air leaks with a lighted candle! Even so, when compared with any other industrial operations, the safety record of nuclear power plants worldwide is excellent. What is more, nuclear power plants can still be built within budget and on time.

The trouble that most countries have in the development of nuclear power has little or nothing to do with the intrinsic efficiency or safety of the process. But these problems still give the industry bad press. As we describe here, there is a nuclear power plant at Bataan in the Philippines which cost some $2.2 billion and took twelve years to complete, but now stands a still and silent witness to the consequences, not of the failure of nuclear energy, but of political intrigue, corruption, and technical lapses. Ordered and built in the Marcos era and completed in 1985, the plant is said to have a number of faults and may never run. The Aquino government alleges that Westinghouse, who got the contract to build the plant, bribed Marcos to get the order and has done sloppy work. The government is seeking restitution—the debate generates a lot of controversy and heat but no electricity or light.

REGULATIONS GALORE!

Both capital and production costs of nuclear power plants have escalated due to technical problems, but a major factor has most certainly been the ever-growing number of regulations that must be complied with. In the wake of Three Mile Island, environmental concerns, fueled by various vocal opposition groups, have generated an immense array of additional and, in some cases, frivolous safety regulations. One of the earliest cancellations came at the beginning of 1984. A half-finished nuclear power station at Marble

Hill, Indiana, was abandoned after some $2.5 billion had been spent. The Marble Hill Plant is thought to be the most expensive nuclear plant project ever to be abandoned in this way. Consisting of two units, one unit was said to be about 60 percent ready, while the other was about a third of the way along. But the electric supplier, the Indiana Public Service Company, said that they did not have the cash resources to finish the project (*Nucleonics Week,* 1984). The cost of complying with additional regulatory safeguards simply became prohibitively expensive, particularly given that their rates were set by the utilities commission and they stood little chance of realizing a break-even return on their investment. They were not the only public utility to find themselves in that position.

The Washington Public Power Supply System undertook an ambitious nuclear power plant building program. Five plants, designed to serve some eighty public utility companies, were originally budgeted at about $4 billion. By early 1982, that figure had soared to $24 billion—six times the original estimates! Subsequently, two of the proposed plants have been cancelled, while work has stopped on two others, leaving only one that might reach completion (*Daily Telegraph,* 1983).

Of course, all this assumes that the management of the design, construction, and operation is effective and efficient. As we have seen earlier, management can be careless and operators negligent. The Tennessee Valley Authority (TVA) had a most ambitious program for the development of nuclear power, but there came a time when they had nine reactors under construction, costing $9 billion, but not one operating. TVA brought in a retired admiral, Steven White, from the nuclear navy and gave him an army of engineers to make things work, but found themselves worse off as a result. White's actions in attempting to get the government to ease safety restrictions came under increasing fire. The admiral was alleged to have a "conflict of interest," to have "violated government pay guidelines," and shown "bad judgement." The sad result for TVA was that they were paying $1 million a day for lost power that they had to purchase elsewhere, all as a result of bad management. The problems that TVA have experienced are too numerous to mention, but in the midst of the congressional hearings against White, one

TVA board member said: "I guess I'll just have to go to the president and say, 'Here are the keys to $15 billion in nukes.'"

WHITHER COST ESTIMATES AND PROJECT MANAGEMENT?

The cost of building nuclear power plants in the U.S. has been skyrocketing. To take a typical example, let us consider the cost of the 1,100 MWe Fermi 2 units being built for Detroit Edison (*Business Week*, 1984). Over the years, the estimates have gone up well over ten-fold. Such a record makes a mockery of cost estimating and project management. The end result of such escalating costs for nuclear power plants is that the coal-fired plants begin to appear competitive once again, even allowing for the inevitable escalation in the price of coal over the lifetime of the plant (Guentcheva and Vira, 1984). Operating data for 1983 in the U.S. indicates that the nuclear and coal electrical generation costs ended up roughly the same, although costs vary widely across a country as large as the United States (Energy, 1985).

There has also been talk of converting some half-finished nuclear plants for coal firing. One cost-benefit analysis on the completion of unfinished plants shows that in most regions of the United States, finishing the units is not the least costly of the various alternatives (Feldman, Bernstein, and Noland, 1988). Another complication is the fact that in many regions there will be no need to increase the provision of power before the year 2000.

Yet it all began so well. David Lilienthal, a giant in the field of nuclear energy, has been closely associated with the development of atomic energy in the United States for more than thirty years. In early 1947, he was appointed head of the newly created (and civilian) Atomic Energy Commission. In his book, *Atomic Energy: A New Start*, he tells of a meeting with President Truman in which the president said to him: "You have the most important thing there is. You must make a blessing of it." Lilienthal (1980) goes on to say, with that remark in his mind:

"Bombs we now have, by the thousands, as do the Russians. But we do not yet have a safe method of producing atomic electricity."

A $5.5 Billion Nuclear Power Plant Sold for One Dollar

The Shoreham nuclear power plant in Long Island, New York represents one of the most infamous examples of the abortive development of a nuclear power plant. Poor project planning, inadequate management troubleshooting, and public relations coupled with public activism and political expediency have all blended to create one of the truly momentous examples of a project disaster. Fearful of nuclear power, Long Islanders have forced LILCO (Long Island Lighting Company) to sell the completed $5.5 billion facility to LIPA (Long Island Power Authority) for the price of one dollar! On top of that (as if that was not enough!), the New York state government (owners of LIPA) has pledged to demolish it, for free (*The Economist*, 1989). While the above sounds like a fairy tale, it is one that must be believed, for it is true indeed! How and why? Simply put, antinuclear activism coupled with the project's own mismanagement were the root causes. Mr. Wasserman, a project opponent, (1988: p. 21) expounds on the nuclear debate from his perspective:

"About a trillion dollars has been squandered on construction of atomic energy plants, and another trillion or so may be spent on attempts to dispose of nuclear waste . . . [thanks to] Three Mile Island . . . and Chernobyl. This spring, opponents of nuclear power were able to claim two huge victories, one at Seabrook and the other at the Shoreham [plant] . . . Though 100 reactors are in operation in the U.S.—and 350 worldwide—the prospects for nuclear power have changed dramatically . . . The question is . . . whether another disaster will occur before alternatives can be brought on line and the last reactor shut down and entombed . . . [The Seabrook and Shoreham plants] represent [an investment of] almost $11 billion . . . more than enough to have pushed . . . a long way toward development of . . . solar power."

Coincidentally and ironically, the sale took place just as the Nuclear Regulatory Commission issued a full power operating license for Shoreham, some five years after the application. Why did popular opinion turn dead set against this project?

The primary issue that derailed Shoreham was safety. In case of a serious accident, the large-scale evacuation of Long Island would be far more difficult than the areas that surround most other nuclear plants. The points of egress from Long Island make it virtually impossible to organize and execute a mass evacuation within a reasonably safe time period. In spite of the obvious limitations with the design and location of the facility, it had its share of proponents. Mr. James Watt, Secretary of Energy at the time, termed the decision to demolish this plant, "Alice in Wonderland." Local consumers were also angered at the promise made to LILCO by Governor Cuomo, that the state would compensate the company for its investment loss by allowing increases in rates—in effect, forcing taxpayers to cover the cost of dismantling the facility.

Governor Mario Cuomo, the man ultimately responsible for killing the project, was quoted as saying: "Shoreham has to die, Shoreham will die." In the process, not only do the citizens face higher rates, but also the possibility of blackouts. Critics of Cuomo's actions have argued all along that the so-called safety issue has been a red herring. They point out that there has not been a nuclear radiation fatality in the U.S. ever since the advent of nuclear power nearly forty years ago. Even the Chernobyl fiasco, the worst ever, which claimed thirty-two lives, resulted in far fewer deaths than occur from coal industry accidents annually in the U.S. Meanwhile, other countries, notably France, Japan, and South Korea have been forging ahead with nuclear power, without the sort of paralysis that has occurred in the U.S. (*Wall Street Journal*, 1988).

Ordered in 1967, the Shoreham plant did not get a construction license until 1973. Building Shoreham took eleven years, but as we now know, building it turned out to be comparatively easier than operating it! In sharp contrast, the Millstone Plant II, across Long Island Sound, ordered the same year as Shoreham, got its construction permit three years earlier, and was completed in five years (instead of eleven). By the time Shoreham was finished, Millstone had already paid for itself through fuel savings of $700 million, against its total cost of $424 million. Meanwhile, what was Shoreham's final cost? Five and a half billion dollars! Among the main reasons for this fiasco are (McCracken, 1988):

- Management failures—Through poor control processes, costs were allowed to skyrocket out of proportion to the work being performed. Further, management had either poor or tepid relations with the various unions and their officials, resulting in numerous slow-downs and delays.
- Heavy-handed regulation—It was impossible for the construction managers to gain any sort of learning curve benefits as the rules were continuously changed in midstream. Safety requirements were routinely scrapped and updated at prohibitively expensive levels, all while construction was ongoing.
- Environmental guerrilla tactics—The facility was preyed upon by a number of environmental groups who made it their life's purpose to delay or stop the construction through demonstrations, acts of sabotage, and continually issued court injunctions and other litigation mechanisms.
- Malevolence of the local governments—Increasingly, the plant was seen as a political "hot potato" in which little good came from supporting its construction and much publicity and exposure came from opposing it. A snowballing effect developed in which more and more local governments formed alliances to block the development of the plant.

Unfortunately, the nuclear power standards were set at almost impossibly high levels, ones which no other industry could ever meet, including the coal-fired power plants which normally emit more radiation than is permissible for nuclear power plants. Apparently, these standards were still not adequate for Governor Cuomo, who upgraded the near impossible standards to "infinitely high" ones! What is the lesson that the American nuclear industry and its opponents can learn from Shoreham? (*The Nation,* 1988: p. 812):

> *"If people organize and fight back they can win, even against the formidable forces . . . Big government . . . Big business . . . Big banks . . . Big Science . . . and Big Media . . ."*

We have tried here to bring you the salient points about Shoreham because of its importance in the present context. But we thought a summary of the final events during the plant's brief existence, as chronicled by the *New York Times* in their annual indices, may help recapitulate:

- October 1991: For over twenty unhappy years, this project has simply generated bills and controversy. About to make history as the first fully licensed nuclear power plant to be dismantled without ever operating competitively.
- November 1991: As Shoreham nuclear plant nears its demise, the district is anticipating 90 percent loss of its revenue in ten years.
- February 1992: Nuclear Regulatory Commission approves transfer of LILCO to Long Island Power Authority for dismantling.
- March 1992: Transfer of this $5.2 billion plant takes place for a nominal one dollar, thus ending a twenty year battle. Plant to operate possibly on natural gas.
- January 1993: LIPA chairman defends agreement with the French company, Cogema, for reprocessing Shoreham's contaminated nuclear material.
- February 1993: Home found for partly spent nuclear fuel; a buyer pays $45 million.

- 1994: Intense debate about Long Island Power Authority's plan to convert what is left of Shoreham nuclear plant into a gas-powered generator.
- 1994: Barge carries last shipment of uranium fuel from Shoreham, another milestone in closing the $5.5 billion reactor that brought the Long Island Lighting Company close to brink of bankruptcy, leaving Long Island with among highest electricity rates in the U.S. (*New York Times,* 1994).

THE $2.2 BILLION NUCLEAR FIASCO

We borrow our section heading from a cover story (Dumaine, 1986) in *Fortune* magazine on the tragicomic story of the Philippines first and only venture into nuclear power. Supplied by Westinghouse, the Philippine power plant (600MW) has been referred to as a "management nightmare." It has never been put into operation amidst controversy that the company bribed Ferdinand Marcos. The company's work was thought to be generally sloppy, with faulty supports, submerged power cables, and broken cooling fans. The plant has been considered ready for operation since June 1984, but even a decade later, it remains bogged down with controversy, corruption charges, and technical defects. It has never been commissioned despite the urgent need for power in an energy-starved country. Present indications are that it may never operate, though it has been costing the Philippines exchequer millions to maintain all these years!

The history of the plant was checkered from the first day. Originally ordered in 1973, the contract was awarded without soliciting competitive bids. In spite of GE's generally positive and well-received presentation, Westinghouse had secretly been awarded the contract a week earlier. This decision is particularly interesting since Westinghouse's bid was estimated to be at least 15 percent higher than GE's bid. The final, agreed upon price for the power plant was $320 million. Although initial projections calculated a construction period of no more than five years, it was not completed until June 1984, at which time the power plant was said to be ready for operation.

The truly intriguing elements of this project concurred with the actual construction and showed increasing evidence of "sweetheart" deals, rampant institutional corruption, payoffs to governmental officials, and fraud on a massive scale. The details that we pass on here have been gleaned from news items reported in the *New York Times* annual indices for the period under consideration.

By 1978, the plant was less than half finished. At the same time, rumors of high level payoffs by Westinghouse had become so persistent that President Ferdinand Marcos ordered a governmental probe of these alleged payoffs. The official target was said to be a Mr. Herminiot Disini, a friend and relation by marriage to Marcos. A handwritten note recovered from Westinghouse files indicated that Disini, referred to simply as the "man from Marcos," served as the go-between, at one point carrying $17 million to bank accounts in Marcos' name. Because there was no hard proof of these transactions, the Philippine government, controlled by Marcos, made no effort to explore the charges and the investigation was quietly shelved.

By 1980, the original projected cost of the power plant ($320 million) had ballooned to $1.1. billion, with no end in sight. An additional $644 million in loans were extended by the U.S. Export-Import bank while the Philippine government opposed any arbitration settlement with Westinghouse. Meanwhile, the work continued.

In 1981, faced with escalating cost overruns, the U.S. Export-Import Bank approved an additional $104 million financial guarantee for the power plant.

Three years of additional labor went into the power plant before it was "officially" completed in June 1984. Unfortunately, inspections revealed a litany of problems, including shoddy workmanship, cracks in the reactor housing and containment dome, poor welds, inadequate cooling, and so forth. At this point, the Philippine government made the decision to hold off on activating the plant until additional follow-up work could bring it up to current safety standards.

With the fall of the Marcos regime, governmental reformers took a hard look at the power plant, probing in particular the earlier investigations of payoffs. In March 1986, former President Marcos was alleged to have received an $80 million payment from Westinghouse ten years previously, for granting this lucrative contract. In April of that same year, the cabinet decided to halt remedial construction and take Westinghouse to court to void the contract, seek damages, and recover their costs.

By May, the Philippines government was in a quandary. They had already spent $2.1 billion on a plant that was essentially useless. In fact, even if it had started up, the safety violations associated with the plant would have ensured that it was a serious safety hazard, to the point of having catastrophic consequences for Manila and outlying provinces.

In 1988, the government of the Philippines filed suit against the two prime contractors, Westinghouse and Burn & Roe.

In March 1992, following intensive talks to settle the matter out of court, the following tentative agreement was reached: Westinghouse would repay $300 million in overcharges and the Philippines government would spend an additional $400 million for repairs and maintenance to the power plant.

In November 1992, the new government of President Ramos rejected the tentative settlement agreement and decided to press its claim in the U.S. court system. Manila hoped to recover damages, also $44 million in litigation expenses and $11 million in maintainance of the (unoperating) plant since its completion in 1985.

The long story of the Philippines nuclear power plant was finally settled in May of 1993, a full *twenty years* after the contracts were originally signed. On that date, a civil jury in the city of Newark, New Jersey cleared Westinghouse of bribery charges. Six months, later the two parties agreed to sit down and negotiate an out of court settlement. The details of that settlement are still pending.

Lest we give the impression that this case is indicative of Westinghouse's business dealings in general, we should point out that Westinghouse has one of the best records of any construction company in the nuclear industry. It has built seventy-one plants, operating in twelve countries, and there has never been a major accident in any of the plants they have built. The company is still seeking business in Egypt, South Korea, Serbia, China, Taiwan, and The Netherlands. Certainly, we can hope that the Philippines case was an aberration. Meanwhile, the power plant on the Bataan peninsula continues to lead a star-crossed existence. It is said that "hundreds of millions of dollars" are required to get the plant working properly; yet in February 1985, joint IAEA and Philippine Atomic Energy Commission experts concluded that the plant meets the "international safety standards followed by twenty-six nations." Where the truth lies appears to depend upon whom one asks.

At the root of the trouble is the suspicion of bribery, but the issue has been clouded by a variety of other matters. A volcano, Mount Natib, looms up behind the power station and the experts cannot agree on whether there is any danger. One expert states: "The fuss about the volcano is all smoke. Chances are zero that it could endanger the plant." Currently, the plant has now been mothballed and a study has been ordered to see whether it can be converted to coal firing. At the same time, interest charges mount and the Philippines continues to face an electricity shortage, with frequent brownouts and blackouts.

Meanwhile, it has been reported that to bring the plant to a condition fit for operation will cost $200 million and take three years. As we noted, an alternative has been suggested to convert the plant to nonnuclear usage. This is expected to cost about $600 million according to estimates of a major American contractor, Fluor Daniel. The term "conversion" is really a misnomer, since it means, in effect, a new diesel-fired facility complete with its own set of generators. It will make large parts of the nuclear plant, namely the reactor and containment shell, redundant and they will have to be leveled and disposed of as junk. In an editorial, the daily *Manila Chronicle* (1993) summed it all up:

> *"The costliest power plant in the world sits in Morong, Bataan . . . this may be a monument to our folly or to our anger."*

AUSTRIA: PLANT COMPLETED, UNFORTUNATELY

Unlike the Philippines example, one of huge cost overruns, fraud, poor construction, and mismanagement, the nuclear plant built in Austria illustrates another risk associated with nuclear development: The environmental and legislative backlash against nuclear power. As such, this case demonstrates the power of what David Cleland (1988) has referred to as "Intervenor" stakeholder groups: those who can play a huge role in the viability of nuclear power plant construction without having any direct input into the construction itself.

Austria's first ever nuclear facility was completed at Zwentendorf in the late 1970s. A model of high technology, the plant experienced few of the construction problems that are often associated with nuclear power stations. Ultimately however, its construction record had nothing to do with its long-term viability. In fact, the plant has never produced a single kilowatt of nuclear-generated energy. Why? Because in 1978, opponents of the plant organized a nationwide referendum against its commissioning that was narrowly approved. Then Chancellor Kresisky staked his political reputation on his pro-nuclear policy and lost at the hands of the voters. Although not technically bound to abide by the results of the referendum, Kresisky acceded to the wishes of the majority and ordered the plant mothballed. At the same time, he ignored the vociferous warnings of members of his own party, including the speaker of Parliament and his own finance minister, that Austria cannot afford to be without nuclear power.

Finally converted on a limited basis to conventional fuel usage, the plant has been a financial drain on the country ever since its original intent was aborted at the eleventh hour. By the 1980s, the plant was costing nearly $73 million a year to operate and was contributing to the pollution problem that continues to plague Central Europe.

Not only has Austria decided to dispense with nuclear power, she wants her neighbors to do likewise on the basis of future risk. For example, Austria has formally requested that the Czechs close down their Soviet-built, two-reactor nuclear power plant and has even offered to supply, at its own expense, the energy equivalent needed to make good on the loss.

The Austrian case is interesting in that it demonstrates the inherent risks in developing nuclear power plants. The risks in this case are not technical in nature but rather, political and social. Nuclear power is a challenge in any sense of the term and the current state of public attitudes regarding nuclear power renders any construction project that much more onerous. It is hardly surprising that the regions where nuclear power are continuing to make inroads (China and the Pacific Rim, in particular) also have a more quiescent population that is less likely to mount the sorts of systematic protests that could halt nuclear power development.

TURNING LEMONS INTO LEMONADE

This is the unusual subtitle of an article (*The Economist*, 1990) that relates the story of the commissioning in April 1990 of the conversion of the first nuclear power station to conventional fuel. It was hailed as a forerunner of many others to follow. Why? The two major accidents, TMI (1979) and Chernobyl (1986) produced such fierce opposition to nuclear power that plans were afoot to convert several abandoned or mothballed nuclear power plants.

At the commissioning of the world's first ever nuclear power station converted to natural gas (one of the conventional fuels), the plant's main owner, Mr. William McCormick, chairman of the CMS Energy Corporation, was so enthusiastic about the project that he devised a snappy marketing slogan:

We are in the business of turning lemons into lemonade!

This was expected to give the company a headstart in a potentially huge market. In a confident prediction of other conversions to follow, McCormick reckoned that $20 billion worth of abandoned and/or mothballed plants may follow suit. Utilities across the States were finding it difficult, even impossible, to complete nuclear power plants even fairly advanced to the finishing line.

The Midland plant was seven-eighths completed when Consumer Power, the operating subsidiary of CMS Energy, stopped work in the face of the growing political opposition and soaring costs, which had already reached $4.1 billion. CMS feared the worst, even bankruptcy, with its share price crashing from $21 to $4. Quick action, conversion to "lemonade," radical surgery on costs, plus approval for increases in electricity rates, have all helped the company, with share price springing back to $32 (March 1990).

Midland went a step further with their "vision" and farsightedness. Hoping that Midland would be the forerunner of several such plants, CMS formed a partnership with six other companies for the conversion and operation of the converted plant. About 60 percent of the nuclear plant's hardware, including the nuclear reactor and everything connected, therefore became redundant.

Other opportunities for conversions abound:

- Zimmer Power Plant (Moscow, on the Ohio river): 96 percent complete as a nuclear facility, then blocked for conversion to coal. Commissioned in 1991.
- Rancho Seco (California): shut down by a referendum, summer 1989.
- Shoreham, NY: detailed in the story above; five years later, there seem to be no signs of conversion, and the plant is slated for "demolition."
- Sweden: Could be a lucrative market for such conversions to meet their commitment to phase out nuclear power. The phase out seems to have since been postponed.

All of the above are examples of the current state of ambiguity and confusion that reigns in the nuclear industry in the West. In contrast with the continued expansion of nuclear power in the Pacific Rim, most of the developed countries today have made what can essentially be described as the politically sensitive decision to limit or phase out nuclear power. As we have noted, the "down-side" risk is just too high for many countries to commit to nuclear power for the long term.

Unfortunately, and perhaps, ironically, that view can also be said to be excessively short-sighted as it presumes a limitless supply of other forms of energy (chiefly coal and oil) as sources of electricity and power. Clearly, this attitude is erroneous; the earth has only limited supplies of fossil fuels and the continued development of the third world coupled with the high usage rates of the West guarantee that these supplies will not last forever. It is not our intent to offer an apology for nuclear power; rather, we simply point out some undeniable facts. The world will need greater and greater amounts of energy that must be tapped from dwindling supplies. Whether or not the idea is palatable, it may be inevitable that we come to make peace with the use of nuclear power, at least until alternative power sources can come on line in quantities great enough to sufficiently ensure our energy for the future.

Fundamentally, the issue that is most directly relevant to project management from these cases is the importance of active stakeholder management. Stakeholders are any group that has an active stake in the outcome of the project. In this context, each of the environmental and community groups fighting the development of nuclear power are important stakeholders of the process. One of the most important messages for project managers to internalize is the need to brainstorm the individuals or groups who can affect successful development, seek them out before the project kicks off, and try to come to some meeting of the minds as to how they can benefit from the completion of the project.

Obviously, in the case of these nuclear power plants, the cards were stacked against the utilities even before they began active stakeholder cooption. The Three Mile Island episode cemented the inherent dangers of nuclear energy in many people's minds, and made future installations an uphill fight from the start. Nevertheless, it is interesting to note how few of these utilities actively worked with stakeholders prior to receiving permission to begin installation. One of the standard approaches for a long time was to ram the new plant down the public's throat, keeping them out of the picture until the plant was ready to go on line. Obviously, by the late 1970s, this strategy no longer worked. But while it was in use, it created a whole generation of opponents of nuclear energy and the "reckless arrogance" of the public utilities.

Project managers must consider the external stakeholder environment when initiating a new project. Not only should they scan those who would be affected by developing the project, but also those who are intended to be future customers for the project's outcomes. A vital question to ask ourselves before we bring our projects too far along is what is the marketplace? Who are the stakeholders, and am I aware of their concerns? Do I treat those concerns legitimately or am I inclined to minimize their anxieties or resistance? The answer to these questions will give project managers an important road map for managing stakeholder relationships—among the most important roles in which they will engage.

This chapter has offered some examples of the current state of the development of nuclear power. Clearly, the problems facing its use are far more than simply managerial; they comprise the public relations aspect, the technical safety components, and the political dimension as well. If nuclear power does make the comeback that many people argue it must, it can do so only when it has gone far enough toward addressing many of these diverse concerns.

References

Business Week. 1984. "Where Utilities and Anti-nuclear Activists Agree," 16 April, p. 185.

Cleland, D.I. 1988. "Project stakeholder management," in D.I. Cleland and W.R. King (eds.), *Project Management Handbook, 2nd ed.,* New York: Van Nostrand Reinhold, pp. 275–301.

Daily Telegraph. 1983. "Whoops, we cannot pay $1,433 million," London, 17 May, p. A1.

Dumaine, B. 1986. "Nuclear scandal shakes the TVA," *Fortune,* 114, 27 October, p.38+

Dumaine, B. 1986. "The $2.2 Billion Nuclear Fiasco," *Fortune,* 114, Sept. 1, pp. 14–22

The Economist. 1989. "Lights off on Long Island," vol. 311, April 29, p. 33.

The Economist. 1990. "Converting nuclear power stations," vol. 314, April 3, pp. 80–85.

Energy. 1985. "Nuclear and Coal Equal in 1983 Generating Costs," vol. 10, No.1, Winter, p. 27.

Feldman, S.L., Bernstein, M.A. and Noland, R.B. 1988. "The costs of completing unfinished US nuclear power plants," *Energy Policy,* June 16, pp. 270–279.

Guentcheva, D. and Vira, J. 1984. "Economics of Nuclear vs. Coal—Big is Beautiful?," *Energy Policy,* December 12, pp. 439–451.

Lilienthal, D.E. 1980. *Atomic Energy: A New Start,* Harper & Row, New York.

Manila Chronicle. (1993).

McCracken, S. 1988. "Impossible Safety Standards," *National Review,* June 24, p. 14.

The Nation. 1988. "A Grass-Roots Triumph to be Duplicated?," vol. 246, no. 23, June 11, pp. 811–812.

New York Times. June 4, 1994, I. 24:3.

Nucleonics Week. 1984. "Marble Hill Owner rejects industry joint venture bid to complete unit," vol. 25, No. 25, 21 June, p. 1.

Wall Street Journal. 1988. "Do Citizens Lose?," May 31, p. A-24.

Wasserman, H. 1988. "Averting Nuclear Power," *Progressive,* vol. 52, no. 8, pp. 20–21.

15

The Channel Tunnel: Is There Light at the End?

INITIALLY ESTIMATED TO COST $7.5 billion with completion scheduled for 1992, the Channel Tunnel ("Chunnel") was finally completed in mid 1994 and opened to traffic four months later. The final price tag was $17.5 billion. And this disaster happened despite the fact that it was a private sector project! Priced at break-even cost for users in order to compete with the ferry, the "Chunnel" seems destined to be a disaster and at best, a "gift" to the next generation. It turns out to be a classic project failure with some valuable lessons. A fascinating story, we bring it to you here.

Author's Note: We clearly recognize that of all the projects studied in this book, this chapter is likely to generate the most reaction based on our assessment of it as a failure. From a technological viewpoint, no one will argue that the "Chunnel" was a masterpiece of work. On the other hand, the project has been saddled with huge losses that continue to plague the investor groups who funded it. Further, in spite of the assurances to the contrary, there is a loud argument made to the effect that its development was never as necessary as its proponents argued. The Chunnel may well be a gift to future generations, but for now, it also represents a financial nightmare.

INTRODUCTION

The "dream" of tunneling under the English Channel between France and Great Britain goes as far back as 1802 when Napoleon considered it and other far-fetched methods for invading England at the height of the continental wars. The dream has flickered and flared ever since, finally becoming a reality in 1994. But the dreamers of the past probably knew more about digging than about politics.

The Channel Tunnel, Eurotunnel, or "Chunnel" as it is called, would certainly have been a reality long ago if only the British were not so worried about such easy passage for possible invaders. To be honest, the feelings on both sides of the channel regarding the establishment of a permanent link were no better! Schemes had floated for decades (indeed, centuries) to create a permanent link between England and the European continent. In fact, one early attempt in 1880 was abandoned two years later due to British fears of an invasion via the access route. Interestingly, the remnants of the 1880 tunnel were actually used as a service tunnel for the existing Chunnel, thus establishing a continuity for over a century and providing an historical as well as physical link.

The Chunnel was expected to change from being a mere hole in the ground to a lucrative, revenue-creating business from its opening day. As we shall see, current projections are far from rosy, estimating that the Eurotunnel will continue to operate in the red, not only through the turn of the century, but probably even well beyond that. Such has been the highly escalated capital cost with an enormous interest burden to service the loan. And the latest news is not encouraging, either. August 1995 tourist traffic figures were disappointing, though performance on the freight hauling end of the business was encouraging. In both cases however, the forecasts have continued to be overly optimistic (*Economic Times*, 1995).

A True Love/Hate Relationship

In order to understand attitudes regarding the Chunnel, it is important to gain some insight into the psyche of both the English and the French, particularly with regard to any sort of permanent connection. When presented with a tunnel scheme in 1858, Prime Minister Lord Palmerston feared that it would (Pedersen, 1994):

> *"shorten a distance we already find too short."*

Historically, the divide between the two nations has been cultural far more than merely geographic. Neighbors with a curious mixture of animosity and friendliness, Britain and France often seem to the rest of the world as two married partners constantly bickering. However, as in the case of domestic relationships, when either party is threatened by an external source, it is usually the other who quickly comes to their aid. Nevertheless, this strange, dialectical relationship has had its effect on any efforts to seriously consider creating a permanent "land bridge" between France and the British Isles. To this day that spirit still pervades, especially in Britain. A poll taken in mid 1994, just after the Chunnel's formal inauguration, revealed that 75 percent of Britons doubted they would ever use it. An earlier survey (Nicholson, 1992) reported that 40 percent of Britain's population would be "afraid to travel by the tunnel." The completion of the project was still about a year away. But even at that time, the outlook was none too bright with a dismal prediction:

> *"It could conceivably fail due to underuse."*

The issue of the prestigious *Economist* (April 30, 1994), on the eve of the formal inauguration of this mega project, had extensive coverage of the event. Three of the five pages covering this project were commissioned (this is an exception indeed for the *Economist)* from Andre Fontaine, a former editor of *Le Monde* and one of Europe's leading commentators on international affairs. Talk of the Chunnel as a gift to the next generation! And yet throughout the material describing the Chunnel as a technical masterpiece, even the authors could not wholly escape from the fundamental differences that the two sides often feel toward each other. Consider, for example, the following quotes that were liberally sprinkled throughout *The Economist*'s pages in an article commemorating this impressive achievement (Fontaine, 1994: pp. 19-20):

> *The war of all wars, the combat of all combats, is the one between France and England. All the rest are mere episodes.*
>
> (Jules Michelet, French historian)

> *Fundamentally, our two countries have always been at war, save when they were allied against a common enemy.*
>
> (General de Gaulle to British ambassador)

> *One might even begin to conclude that the traditions and contradictions which separated, even opposed the British and French peoples for so long were now disappearing.*
>
> (Fontaine)

> *You may be sure, General, that if we ever have to choose between Europe and the open sea, it is the open sea we would choose.*
>
> (Churchill to de Gaulle on the eve of the Normandy landings.)

THE PROJECT: TUNNELING UNDER THE CHANNEL

The notion of a single "Channel tunnel" is something of a misnomer. In actuality, the Channel tunnel is really a trio of tunnels built in close proximity to each other. Two of the tunnels were designed to carry high-speed trains. Supporting shafts and access ways had to be constructed along the entire length as well. The technical challenges associated with a project as massive as this were truly daunting. Workers had to contend with drilling tunnel shafts while sometimes working over 350 feet below sea level, maintaining fixed reference points to drillers on the other side of the channel working toward them, and remaining within stationary and stable geological strata to ensure the tunnels would not collapse. The tunnel was constructed through a chalk marl stratum, situated well below the seabed.

Other technical marvels of this project include the construction of two crossover chambers, each at the one-third point from the respective terminals. These chambers allow the trains to change tracks once underwater, for maintenance purposes. At over 480 feet long, 58 feet wide, and 33 feet high, these crossover chambers are the largest underwater caverns in the world. In order to reach the correct geological depth to start work, a huge circular shaft was sunk in order to gain access to the point where the tunnel boring machines would be run. To give the reader an idea of the logistics of this process, the shaft, with a diameter of over 165 feet and a depth of 225 feet was large enough to hold the Arc de Triomphe in Paris.

The primary contractor, Transmarche-Link (TML), is a consortium of ten major construction companies (five British and five French) that was responsible for designing, constructing, and commissioning the tunnel. Transmarche-Link is a huge organization which, at the height of tunnel construction, numbered more than 14,500 personnel (See table on following page). Average expenditures were running about $5.5 million per day. The contractor's total span of responsibilities was enormous. They were responsible for digging out the three tunnels (a total length of over 170 kilometers of drilling and excavating work) from Folkestone, England to Coquelles, France. Further, as two of the three tunnels were designed to carry high-speed passenger rail lines between the two points, TML was also responsible for designing and constructing the locomotives and wagons, the rail system on which they run, the catenary that brings power to them, the signaling that guides them, and the large number of ancillary supporting systems essential to a state-of-the-art railway (Lemly, 1992).

List of Member Companies for Transmarche-Link

U.K.

Balfour Beatty Construction, Ltd.

Costain Civil Engineering, Ltd.

Tarmac Construction, Ltd.

Taylor Woodrow Construction Holdings, Ltd.

Wimpey Major Projects, Ltd.

France

Bouygues S.A.

Lyonnaise des Eaux Dumez

Societe Auxiliare d'Entreprises S.A.

Societe Generale d'Entreprises S.A.

Spie Batignolles S.A.

The contractor's work did not end there, however. At each end of the Eurotunnel system, huge terminals had to be designed and constructed by TML. TML was responsible for laying tracks, building bridges and service buildings, and constructing access ramps, platforms, and terminal buildings. Over 20,000 lighting fixtures were installed in the tunnel system, with almost 900 miles of power-carrying cable strung. All in all, the Chunnel represents a huge undertaking, one that even at the outset was estimated to cost $7.5 billion.

In March 1986, the Treaty of Canterbury established a concession agreement under which the tunnel project could commence. Its most compelling features were its guarantee against political cancellation and the grant of a concession to the private promoter group. That concession, which is scheduled to last for fifty-five years, is a basic agreement which made the Chunnel's construction possible, although, as we will see, it opened the floodgates to a number of contractual complications down the road.

Once drilling commenced, the two sides of the consortium

(Transmarche on the French side and Translink on the English) began drilling while using a sophisticated global positioning system and laser guidance to ensure that even though they started at different points, the two ends of the tunnel would meet exactly, much as the transcontinental railroad across the United States had met at Promontory, Utah. Obviously, directional flaws 300 feet below sea level would have been far more disastrous and expensive to fix than an incorrect railroad junction. Work proceeded smoothly and the first service tunnel linkup was completed on December 1, 1990—exactly on schedule. The two main track-bearing tunnels were completed in May and June of 1991, again running several weeks ahead of schedule. All in all, it was a tremendous technical achievement.

INAUGURATION WITH GREAT FANFARE

On May 7, 1994, the inauguration of the Chunnel was held. Although this ceremony marked the formal "completion" of the tunnel, in practicality it would not be ready for another four months. Nevertheless, this occasion gave rise to a number of euphoric demonstrations by leaders on both sides of the English Channel, including:

> *French President Mitterand: "After two centuries of delays, today is the conclusion."*

> *Elizabeth II, Queen of England: Thanks to engineers and entrepreneurs who have "rejoined what nature had separated for some 4 million years."*

The Economist's verdict was equally positive, though shaded by some of the practical clouds the Chunnel had left on the horizon, reporting on April 30, 1994: "It's a gift from its financiers for which future generations will be grateful."

The Hole May Remain a "Hole"

A miracle of engineering in our times, the Chunnel's very existence was threatened as early as 1990. Faced with serious cost overruns, the international banks were reported to have declined a request for more money and the future of the project was stated to be in doubt. Eventually, additional funding was forthcoming, but the hesitation of the financing consortium sent a small quiver of concern through the contracting organization. Nevertheless, for better or for worse, the Chunnel is now there for all to see and use, if they will! As *The Economist* (1990: p. 51) notes:

> *"As a business it will be losing at least till 2000 and perhaps well beyond—thanks to the likely price war between existing ferry service and Eurotunnel. There is already surplus capacity in the former and Eurotunnel has more than doubled the total capacity."*

A year after its official inauguration and six months after the Chunnel opened to passengers, it faced crushing interest burden and financial collapse. The Eurotunnel must satisfy the bank lenders' concerns about the dire need for financial restructuring (Stevenson, 1995). Another source (Palmer, 1995) talks of Eurotunnel facing a financial crisis amidst a load of debt and falling stock prices. The financial projections consistently show shortfalls, as the engineering marvel digs deeper. Perhaps, the view of *The Economist* must be taken seriously: Eurotunnel's financial frailty has been obvious now for some time and in early April, its chairman, Sir Alastar Morton, feared that the company was close to bankruptcy.

As this book is being written, the financial situation of the Eurotunnel appears dismal and collapse may be imminent unless operating revenues pick up. The loss in 1994 was stated to be 600 million pounds and the revenues were too meager to even service the heavy debt payments through 1995. This hole in the ground may remain a "hole" as far as the Eurotunnel's bottom line projections are concerned until at least the year 2000, and quite possibly even beyond. Current interest charges on debt are running at $750 million a year. During construction, the contractors' chief worries were time and cost overruns, which were more than double the original estimates. Unfortunately, after commissioning, the worry continues to be revenue falling far short of forecasts by a similar factor.

In the most recent updates on the project, the Eurotunnel is reported to have defaulted on loans worth 8 billion pounds ($12.4 billion), unable to service its debt due to a lack of revenues/profits. Current interest payments on the order of 700 million pounds are also due (Dyer and Buchan, 1995). An editorial title in the *Financial Times,* "No End of Eurotunnel" (p. 15) speaks for itself. On the same page, another article by John Gapper (1995), makes the valid observation that the Eurotunnel is relying on the fact that banks have little to gain from "pulling the plug," and will have to go along with restructuring the loans.

Meanwhile, in the wake of the Eurotunnel group admitting their inability to cover debt payments, share price in the Eurotunnel has nose-dived to an all-time low. Starting with an index of 100 in 1988, the stock reached an all-time high of 200 in 1989, before hovering around 100 during the years from 1990–94. Since the tunnel's inauguration and opening, the stock has been in freefall decline, with a September 1995 price around 40–50. According to a news item in the same issue of the *Financial Times,* Eurotunnel is preparing to sue the contractors TML (Transmarche Link) for hundreds of millions of dollars. According to terms of the contract, TML were to provide a system which "must be fit for the purpose," but tests have revealed that many of the tunnel's features still do not work. Other pending claims include a suit against the railway operators for 2.3 million pounds which is currently in arbitration with the International Chamber of Commerce. Even successful settlement of these suits, though an immediate windfall, would not be sufficiently large nor immediate enough to bail the company out of their current financial trouble.

Sir Alastar Morton, co-chairman of the Eurotunnel, blames the company's troubles on the way the project was set up by the British and French governments and their financial and legal advisors. He argues that although the project was privately funded, it would be naive to assume that the Eurotunnel could ever operate independently of political pressures on both sides of the Channel. These pressures, with the resulting bureaucratic and public relations expediencies, interfered with the Eurotunnel's ability to bring the project to fruition at an earlier point, with less bottom line cost to shareholders. In effect, according to Sir Alastar, the real problem with this project was the lack of:

> *"clients driving it forward with a vision of what the operator needed to have."*

Legal advice alone has already cost the company over 40 million pounds. There is a consortium of 225 banks involved in the syndicated loan and the company is being charged a fee for each bank to ask the same legal questions! It is clear that financial restructuring is the only course and this will have to be done during the period 1995–2000, but this action too is pregnant with legal fees.

WHITHER MARKET?

The total market for the Eurotunnel is projected to double by 2013. The company expected to get a third of the passenger traffic and a fifth of the freight traffic (the main competitors being ferry and air), but this has proved to be overoptimistic. For example, the ferry service's natural reaction to opening the tunnel has been to drop their own fares to a point where choosing the Eurotunnel has become an expensive alternative. In addition to the uncertainty of future fares and freight rates, the competition is keen between the three modes—air, train, and ferry.

> *"Eurotunnel's redemption may well lie in fuller capacity utilization and this calls for a massive fare reduction to undercut ferries. After all, despite the much lower fares, ferry offers a "quality" ride especially for the families going on holiday. Who would pay four times as much for a dull ride sitting in their own car in a closed dark room?" (The Economist, 1994: pp. 71-72)*

In an effort to redress some of the consumer and public criticism of the tunnel project, its backers have sought to retarget their message away from its financial woes, aiming instead to focus on its social benefits. Realizing that the Eurotunnel may stay in the "red" well into the early part of the next century, it has been hailed as a "gift" to future generations (*The Economist*, 1994). The moral of this project for its investors is that the early bird often catches a can of worms! Ballooning costs and delays in commissioning have flattened revenue projections and hence, forced the Eurotunnel into a series of new capital raising exercises such as the recent restructuring of its debt.

The privately financed Chunnel does, however, compare favorably with another Anglo-French collaboration, the Concorde. As we have noted in a previous case, the Concorde was built at a cost of $10 billion in today's money, albeit through the taxpayer's expense. Further, the cost overrun was seven-fold compared to two-fold for Chunnel. In both cases, however, there was undue optimism. The total sales potential of Concorde was seen as a few hundred, but thanks to the enormous cost overrun, the customers disappeared and ultimately only a dozen were manufactured and sold "internally" to Air France and British Airways.

Another unforeseen side effect of the Chunnel is that it seems to have dimmed the enthusiasm and craze for private sector mechanisms and their inherent efficiencies. Graham Corbett, financial director of Eurotunnel, the London-based consortium of investment firms that financed this mega project, says that despite the seeming failure of the Chunnel (Guteri, 1994: p. 115):

> *"Governments of all hues will continue to look for new ways of incorporating private sector funds for public works projects."*

Of course, there are failures in both public and private sectors. Hence, we should not blame one sector or the other exclusively on the basis of their record on just one project. As we will see later, though plenty of blame for this project failure may lie with the contractors, the owner (Eurotunnel) must also share the blame. Locking themselves into a faulty contract, in which far too much flexibility was given to the contractor, Eurotunnel sowed the seeds of its present day financial woes with unsound oversight practices.

EVEN THE PRIVATE SECTOR CAN FAIL

In the absence of the government backing of this mega project for over a century, a sort of metacorporation stepped in with a new challenge. It consisted of bankers, contractors, operators, and shareholders. Unlike a public sector project, everything, including delays and cost overruns, was totally transparent; that is, none of the costs could be hidden or written off against other budget items as in the case of government-sponsored work. Of course, the project had the full blessing of both the British and the French governments, who had granted the contractors a fifty-five year "concession" for the tunnel, giving the promoters exclusive rights to develop and exploit the Chunnel. This led to the birth of a separate entity—Eurotunnel—to assume full responsibility for the project, including both the profits and concomitant risks.

Unfortunately, in their enthusiasm to get the project under way, Eurotunnel entered into a faulty contract with the construction companies under the umbrella of Transmarche Link (TML) for a lumpsum turnkey payment. Why faulty? One clear problem was that the contract had no penalty clauses or similar provisions to give Eurotunnel the requisite enforcement "teeth" just in case. At one point, several years into the project, an executive in one of the banks in the original consortium saw the problems that came from lack of oversight and noted:

> *"The builders had written themselves an open-ended contract, and construction was getting out of hand."*

The result was enormous time and cost overruns, with the final cost over twice the estimated expense. To complete the project within the stipulated period of seven years, contractors went full steam ahead even though most of the tunnel's rail and signal systems and trains were in the preliminary stages of design. Further, up front seed money was used to purchase expensive digging equipment. The result was a case of frenetic activity at the project site, perhaps intended to impress political factions associated with the project. Professor Roger Vickerman of regional and transport economics at the University of Kent, noted:

> *"Contractors were too happy to rush off with their spades, and the banks were too happy to rush off with their checkbooks, because they all were eager to make profits."*

This violates a golden principle of project management: Never start site work too early. The activity here happened because the contractors, in their eagerness to spend money as fast as possible, ignored the feasibility and planning activities so important in the up-front stage of project development. The problems were not technical, but rather, stemmed from an unwillingness of all concerned parties to carefully plan the steps in the development. Consequently, two things happened: First, contractors spent a great deal of money without first ensuring that there would be little duplication of effort and wasted activity. Second, in their desire to show dramatic results from the "damn the torpedoes" mentality, design errors began to crop up in the project. For example:

- Shafts were too narrow to carry parts for the tunnel-boring machines
- Tunnel wall linings leaked
- Train carriages did not meet safety requirements

Because of the delays they entailed, all these problems had serious financial repercussions, adding to the substantial overruns. Further, in the absence of design details, the contracts with TML simply specified performance criteria rather than precise blueprints and construction methods. This left numerous gray areas for current and future disputes.

NOW, WHAT TO DO WITH IT?

On the eve of the Chunnel's opening, a leading business magazine reported (*Forbes*, 1994: p. 146):

> *"Now comes the hardest job of all: convincing people to use it!"*

In 1993, about 2.5 million cars crossed the channel by ferry. The Eurotunnel hopes to attract 60 percent of this traffic while charging four times the ferry trip. Their chief selling point is time, comparing their thirty-six minute trip to the hour and a half it takes to travel by ferry. When given the option of relaxing with duty-free beverages on the deck of a ship crossing a scenic waterway or remaining seated in the cramped and stifling darkness, it is hardly surprising to realize that Eurotunnel is faced with a serious marketing problem! The obvious question is how the Eurotunnel, even if it *is* one of the wonders of the modern world, is going to attract people willing to pay extra money to save an hour in an unpleasant environment. Ultimately, more and more potential customers are asking the fundamental question: Is an hour worth the extra hassle?

But this engineering marvel does qualify as a major boon for business travelers. Flying is a bit faster if passengers do not check luggage and there are no delays due to traffic. A train using the Eurotunnel offers a quality ride, with seats equipped with fold-out tables, laptop computer outlets, room to spread out papers, and even room for four-person conferences. The pampering associated with the train, especially in first class, compares favorably with that on an airplane—at a somewhat lower cost. The third alternative, using the ferry, takes nearly twice as long as the train and air travel, but costs about half of either. The price war between ferry, air, and train will favor the customer, but may not do much good to the balance sheet of the operators (Roy, 1994).

Eurotunnel has been a financial nightmare, as this case has noted above. Costs continued to balloon throughout the project's development. The escalating cost in terms of dollars (Toy and

Dwyer, 1993) shows the problems faced by financiers in keeping the project afloat:

November 1987	$7.2 billion
Mid-1988	11.5 billion
October 1991	12.0 billion
April 1992	13.0 billion
April 1993	13.5 billion
Final figure	17.5 billion

Although the project is in its final stages, there is still some last completion work to perform. This work is proceeding far more slowly than Eurotunnel would hope. The cause of the latest delay is litigation between the owners (Eurotunnel) and the contractors. In an effort to settle outstanding claims, Eurotunnel has offered a final payout of $1.5 billion. The contractors are holding firm to their claims for $2.4 billion. Because both sides are refusing to budge, the builders have slowed down the pace of this final work, further delaying the project with additional cost overruns. Promoters continue to hope against hopes, seeing Eurotunnel as the opposite of the Euro-Disney: Easy to build, hard to make profitable. One point is abundantly clear: if Eurotunnel fails, there may be no takers for similar mega projects in the private sector, perhaps for decades to come.

LESSONS FOR MANAGERS

The Eurotunnel project offers some important lessons for project managers, particularly in the area of preproject feasibility studies. No one will deny the impressive technical accomplishment that is represented by the Chunnel. Using the most modern earth-moving and excavation equipment, state-of-the-art designs and materials, and modern siting equipment, engineers on both sides of the English Channel have achieved a wondrous result, creating the first permanent "bridge" between the two countries. We suspect that as a case study in civil engineering and long-tunnel boring, the Chunnel will become required reading in engineering classes well into the future.

Alas, if the story of the Channel tunnel were merely that of a technical challenge met and conquered, this case would have no place in our present book. However, the Chunnel story is not simply technical in nature, nor is its failure due to its technical component any more than the Edsel's failure because of technical lapses in its design. Both cases are fundamental examples of projects that were kicked off with less than adequate feasibility study and only a chimerical image of the potential market. The Chunnel, designed to make travel easier between England and France, offers inflated fares and unpleasant travel conditions in exchange for saving less than one hour of travel time. One is forced to wonder at what point the tunnel's proponents began to ask such obvious questions as who their market was and by what evidence did they assume that they understood it.

The developers only deepened their financial hole with the contracts that they offered to Transmarche Link (TML), which failed to stipulate any penalty clauses or any reasons why the primary contractors should be required to show due diligence in keeping costs down and schedules on target. Certainly, the technical challenges of the project made it difficult to accurately estimate the total time of completion, at least initially. However, after some time had elapsed, the excavation engineers were able to offer fairly accurate projections of future performance. Nowhere in the contracts, however, was any mention given to requiring penalties in exchange for slower or more expensive performance. Consequently, the Chunnel's developers willingly sacrificed their more effective weapon in overseeing the speed and expense of development. That sacrifice had an unintended penalty—because of litigation in settling claims, the Chunnel could not open in time to capitalize on the lucrative summer cross-Channel traffic. In delaying those crucial four months, the Eurotunnel lost an important source of revenue.

Finally, from the start, the Chunnel has been a huge financial drain on the private investors and the consortium of banks that put up the capital to underwrite the project. Even if it were to have come in on budget, its price tag of $7.5 billion would have strapped investors, particularly in light of the debacle of trying to use current, tiny revenue streams to finance and pay down the debt. With a final total cost of over $15 billion, the project has risen to the point where even its strongest backers admit that it is unlikely to see profitability within the next ten to twenty *years.* Restructured debt and falling passenger traffic have both conspired to drive the share price for the Eurotunnel down to less than a quarter of its original list price. From any perspective, Eurotunnel has been a tremendous financial burden, with no positive end in sight.

The Chunnel will continue to garner headlines that pay tribute to the marvelous technical achievement it truly represents. There is, however, a far darker reality that lurks behind these headlines. The Chunnel will continue to be a financial drain well into the future, obviating its usefulness, particularly to a society in which many individuals believe that the distance between themselves and France is already too close!

References

Dyer, G. and Buchan, D. 1995. "Defaults on loans," *Financial Times,* Sept. 15, p. 1.

Economic Times. 1995. "Eurotunnel tourist traffic yet to pick up," Sept. 6, p. A-10.

The Economist. 1990. "Channel tunnel: still soaring a circle," vol. 316, August 28, pp. 51–2.

The Economist. 1994a. "Chunnel vision," vol. 324, April 30, pp. 12–13.

The Economist. 1994. "Eurotunnel—A stormy passage to profit," vol. 324, April 30, pp. 71–72.

Fontaine, A. 1994. "So near, and yet so far," *The Economist,* vol. 324, April 30, pp. 19–22.

Forbes. 1994. "Chunnel vision," vol. 153, p. 146.

Gapper, J. 1995. "Sang-Froid in the Face of Adversity," *Financial Times,* p. 15.

Guteri, F. 1994. "Paying for Light at the End of the Chunnel," *Scientific American,* vol. 270, February, p. 115.

Lemley, J.K. 1992. "The Channel Tunnel: Creating a modern wonder-of-the-world," *pmNetwork,* vol. 6 (7), pp. 14–32.

Nicolson, N. 1992. "Tunnel Vision," *Spectator,* vol 269, December 5, p. 42.

Palmer, J. 1995. "Follow-up—Bottomless Pit," *Barrons,* April 17, p. 12.

Pedersen, D. 1994. "Why do Brits hate the Chunnel?," *Newsweek,* May 18, p. 43.

Roy, S. 1994. "Luxury, Calm, and Speed: It's the Chunnel Train," *Business Week,* Nov. 14, p. 143.

Stevenson, R.W. 1995. "Dimmer Light," *New York Times,* D/1:2

Toy, S. and Dwyer, P. 1993. "Light at the end of the chunnel?," *Business Week,* July 26, p. 21.

16

The Xerox Alto: Death by Neglect

WHEN THE HISTORY of the phenomenal success of the personal computer is written years from now, several corporate names are likely to stand out. Innovative companies such as IBM, Apple, Lotus, and Microsoft conjure images of a period of dynamic technological growth in the U.S., amounting almost literally to a wave that swept a powerful new technology into the average person's home, giving everyone instantaneous access to the wonders of the information age. By the mid 1990s, personal computers have become such a part of our daily routine that it is difficult to consider how work was done prior to their arrival.

Along with the computers themselves, our daily lexicon has expanded to include the terminology of computers. Words like "mouse," "local area networks," "laser printers," and "ethernet" are generally understood and have, in fact, become required learning for new generations of college students entering the workforce. Indeed, the question is not whether we will be affected by computers in our work and home lives, but rather, what, if any, are the boundaries of this seemingly limitless technology.

Personal computers have not simply been a technological phenomenon, but a financial one as well. The buying public spends billions of dollars each year on new, more advanced computers, software, and peripheral equipment, making the computer industry perennially one of the most lucrative in the world and creating millionaires of a generation of young visionaries who recognized the early potential of the industry and created their own rules as they went.

One name that will never appear on anyone's list of giants of the personal computer industry is the Xerox Corporation. Of course, their tragedy lies not in simply missing a major financial and technological wave. Companies fail to recognize strategic opportunities all the time. The true tragedy of Xerox is that they did recognize the opportunities that existed, were perfectly positioned to capitalize on the growth of personal computers, and yet, did nothing. The Alto story represents a classic case of technological brilliance coupled with managerial shortsightedness and corporate politics, all destined to squander a dominant initial position in the personal computer field. This chapter tells the story of how a company that developed the first workable mouse, the first local area network, the first laser printer technology, and the first user-friendly word processing software (all at least *five years* before their competitors matched them) ended up sitting on the sidelines while the personal computer wave passed them by.

The history of the Alto can be broken down roughly into two distinct phases: The project development phase, which was essentially a technical and project development triumph, and the commercialization phase, which was a disaster. As such, it is important to note that project management, and especially project success, cannot be measured by one metric alone. The characterization of a project as "successful" if it is badly managed and a large financial drain on the developing organization is obviously wrong. However, the converse is also true: A project that was brilliantly managed and developed, but poorly received in the marketplace does the company no good either. Research and development are twin pillars. One without the other often signals a sure recipe for disaster.

Up until the development of the Alto, almost all computers operated exclusively in batch format. The user would create a program, get computer time at some point in the future, run it, and review the results. There was no opportunity for interaction with the system itself in a real-time or near real-time mode. Pioneers like J.C.R. Licklider had coined the term "interactivity" specifically to promote the notion that a computer can be used in a hands-on manner.

DEVELOPING THE ALTO

In 1968, as Peter McColough succeeded Joe Wilson as president of Xerox, the corporation stood poised on a pinnacle the likes of which have rarely been seen in American business (Dessauer, 1971). Due to the incredible success of its model 914 office photocopier, the company had enjoyed an unabated string of record breaking years in which sales revenues and profits rose to ever higher levels. With over 95 percent of the market for copiers, they had just posted record performance figures yet again, earning profits of $138 million on revenues of $1.125 billion. As he assumed the reins, McColough's stated intention was to make Xerox a player in the integrated office systems market, including computers. He recognized that a dependence on one product, the copier, was dangerous in the long run. Toward that end, Jack Goldman was appointed head of research at Xerox and given carte blanche to turn Xerox into a world-class company by repeating the success of the photocopier in other arenas. McColough's directive had been short and to the point: Find ways to hit another home run.

In 1969, Peter McColough spent over $900 million to acquire Scientific Data Systems in an effort to get Xerox into the computer industry prior to a move to go head to head with IBM. Up until this point, SDS was a niche player, selling its hardware to scientific firms that wrote their own programs. SDS did not need to spend huge amounts of money providing the operating software for their systems. They were profitable, but not excessively so. Revenues for 1969 were $100 million and profit stood at $10 million. To become a player in the computer market, Xerox paid over ninety-nine times SDS's earnings for the company.

McColough announced in March of 1970 that Xerox would spend over $100 million on the creation of a world-class research center. The Palo Alto Research Center (PARC) was born. Xerox purposely built it away from their East Coast headquarters to give it an independent feel. One of the stated goals of PARC was to serve as a source of pure research for SDS, funneling inventions and innovations into the SDS organization. From its initiation, PARC sent storm clouds throughout the organization, as no one was quite sure what would actually come from its efforts. Further, the one organization PARC's research was expected to directly support—SDS—really did not want anything to do with them, particularly when they found that they would have no direct control over the research agenda coming out of the center. They could not understand the reasoning that would allow PARC to simply work on unfocused pure research when they had their own specific needs.

One of Jack Goldman's first acts was to appoint George Pake, a physicist and former college administrator, to head up PARC. Pake was chosen not for his understanding of computers and computer technology (which was limited), but for his clearsighted ability to pick and choose from among several projects, allocating resources where they could do most good. Pake was the quintessential manager of technical personnel. He had strong technical credentials, but could also recognize the abilities of his subordinates and interact with them without becoming overbearing.

At the same time, PARC was divided into three independent laboratories: General Science Laboratory (GSL)—intended to conduct basic science research, Systems Science Laboratory (SSL)—which had a broad mandate to work on a variety of statistical, mathematical, operations, engineering, and information systems, and Computer Science Laboratory (CSL)—devoted to the study of computer systems.

Throughout the period of 1970–1971, Xerox (principally in the persona of Pake and Goldman) raided universities and rival organizations to assemble a world-class collection of computer and electrical engineering talent. Robert Taylor (although not heavily technical in terms of his credentials, he knew how to work with and direct technical people), Alan Kay (whose FLEX interactive computing system was considered by many to be the forerunner of the modern personal computer), Butler Lampson (one of the earliest and most respected computer scientists), Jim Mitchell, and Chuck Thacker were some of their key acquisitions. Most were lured to PARC with the simplest of offers: The freedom to put into practice their dreams for a new generation of interactive personal computers.

The CSL was a stand-out department from the beginning, insisting on employing a flat organizational structure with little or no impeding hierarchy and management layers. All computer scientists and other personnel (ultimately totaling fifty people) had direct access to Bob Taylor, a process he encouraged. Taylor also demanded regular meetings to serve as a cross-fertilization for the researchers working on various projects. Each week one project leader would present their status, including problems and current solutions, to be quizzed by other members of CSL. Taylor also made one requirement of his scientists, "We use what we build." The lab would not turn out demos that had to be debugged later. Anything built had to survive the continuous usage by the lab members themselves.

In 1972, CSL proposed their first major effort, the Dynabook. The original Dynabook, the forerunner of personal computers, was conceived to be the size of a notebook, portable, and capable of holding an immense quantity of information available for instant retrieval. Although the project was shot down by PARC's management staff in the fall of 1972, it planted the seeds for CSL's next venture, the Alto. In the original proposal, the Alto was described to be almost as powerful as current minicomputers, with a sophisticated "bit map" screen display, and the flexibility to be placed on anyone's desk for instant interactivity—the world's first personal computer.

After several months work, Thacker's team finished the first

Alto in April 1973. Within months, the lab was using forty. Following Taylor's dictum, the machines were put through stringent testing in the laboratory. Once they were operating effectively, they formed the basis for additional breakthroughs in the areas of software and connectivity. CSL developed application software for word processing, the local area network (dubbed "Ethernet") for linking PCs together, and, using and expanding on Xerox copier technology, the first workable laser printer.

In 1974, Ginn & Co., a New England based textbook publisher owned by Xerox, asked for a word processing package that would enable them to perform on-screen editing. By early 1975, PARC had developed Gypsy, the first on-screen text editing and word processing package easy enough to be learned by novices after only a couple of hours practice. In the period of a few short years, scientists at PARC had accomplished what no one else in the industry had envisioned—the practical marriage of a radical innovation in computer hardware (the Alto personal computer) with user-friendly word processing software. By mid 1975, Xerox stood on the brink of dominance in the emerging field of personal computing.

How the Wheels Fell Off: From the Alto to Oblivion

If the story of the development of the Xerox Alto had ended at this point, we could reflect back on the successful results of a dynamic research team's performance in introducing revolutionary and useful technology for the office and home. But the story of the Alto does not end with its R&D effortse. The true tragedy of this case is the manner in which Xerox managed to squander its enormous lead in the personal computing industry over the next several years. If nothing else, the balance of this case should teach the idea that product development is only half the struggle, particularly if the organization is not committed to subsequent product introduction. The Alto represented a golden opportunity for an organization with the willingness to take additional risks, to cut through the miles of bureaucratic red tape and political obstructionism. Unfortunately, Xerox would not prove to be the organization that could exploit the work of its own research group.

Certainly, with the exception of IBM, no organization occupied a better position to profit from the new personal computing technology than did Xerox in 1975. In fact, a *Business Week* article in June of that year predicted that Xerox and IBM were destined to be the big winners in the field of office automation. The article cited Xerox's marketing organization, its current dominant position in the office copier market, and its deep pockets following years of profitability. Xerox had not only the ideas, but the modes of introducing those ideas to the office environment, a combination that seemed to promise nothing but future success.

Unfortunately, within Xerox, other forces were at work, distracting the senior management at precisely the wrong moment to encourage the risk necessary to take advantage of their technological lead. Since 1972, the federal government had been in antitrust litigation with Xerox. The claim was that Xerox had a monopoly on the copier market, a position that they encouraged through predatory pricing and bullying their customers. Xerox, in the form of its president, Peter McColough, indignantly denied the allegations, particularly when the government insisted that they divest themselves of their foreign subsidiaries and stop their leasing policies which essentially kept customers tied to them through leases rather than being able to purchase the machines outright. Even when the government and Xerox came to an understanding in 1975, its chief rivals, including IBM and SCM, continued to push their own private lawsuits in federal court.

A second factor that contributed to the sudden caution with which Xerox management approached strategic decision-making was the recession of 1974-1975. Following a decade of phenomenal growth, Xerox had become a hugely bloated organization by 1975. With profits falling during the economic downturn, they discovered that their internal cost of operations was increasing astronomically. By the end of 1975, the cost of interest on debt, payroll, and materials accounted for 70 percent of revenues, greatly weakening the bottom line (Smith and Alexander, 1988). In that year, McColough announced the unthinkable: Xerox earnings per share were expected to decline. The stock plunged from a high of $179 a share in 1972 to less than $50 a share. Wall Street's confidence in Xerox had been seriously shaken and the top management team began devoting more and more of its attention to bolstering the bottom line, while paying less attention to PARC.

The final impediment to concerted and swift action on the Alto came in the form of continued financial difficulties with its computer subsidiary, SDS. All along, McColough had expected Scientific Data Systems to serve as its wedge into the profitable computer industry. Unfortunately, Xerox's president had fundamentally mispositioned it in the marketplace to compete directly with IBM in a field that "Big Blue" had long dominated. SDS had been successful in the past precisely because it was a niche player, competing in the narrow field of scientific hardware systems and avoiding full entry in office automation. McColough's miscalculation robbed SDS of its strength and pushed it into a game for which it was not prepared: Going head to head with IBM.

The results were predictable: by 1975, SDS had long since lost its profitability and instead, served as a drag anchor on Xerox's overall earnings. Since first attempting to enter the competition for mainframe computers, Xerox had lost $180 million (Smith and Alexander, 1988). In 1975, McColough, now chairman of the board, decided to write off their investment in SDS when an in-house study group concluded that it would be impossible to find a buyer. Xerox's total loss was $1.3 billion.

Far more than the financial loss from SDS, its impact continued to cause repercussions in the corporate psyche at Xerox. Whereas their stated goal for the past seven years had been to become a major player in the office automation field (a cause for which they had spent millions in developing PARC), the utter failure of their venture into computing with SDS caused the top corporate officers to pause and reflect on the direction they were headed. While still committed to the kind of research being conducted at PARC, it was perhaps inevitable that continued losses in their mainframe computer division put the brakes on other burgeoning efforts, including the Alto.

As the head of Xerox, McColough had made no secret of his desire to expand their core business in copiers by making a move into integrated office systems. Nevertheless, at this critical juncture in their history, as they were to redefine the nature of office computing through their initial research into the Alto personal computer, McColough made a move that would have serious and far-reaching ramifications. He promoted Archie McCardell, his group vice president for finance since 1966, to president. At the time, the move seemed a logical one, aimed at putting Xerox's financial house in order. McCardell, a former head of finance at Ford Corporation, was by nature an analyst who tended to make all operating decisions on the basis of their immediate impact on the bottom line. While no one would accuse him of narrow-mindedness, his manner and focus quickly put him at odds with something as nebulous and ill-defined as PARC. Research purely for the sake of research made little sense to his methodical nature.

A second problem emerged in the wake of the demise of Scientific Data Systems. As originally developed, the plan for PARC research was to provide a steady source of innovation for SDS future products. That goal had obviously been superseded, but by what? In effect, PARC research scientists were left without clear direction as to where they were expected to head. Research was continuing apace, but no outlet was earmarked for these developments. At a time of severe internal analysis of Xerox's strategy for the future, no one could clearly articulate the role that PARC was to play nor the types of products they could produce that would do the most good in helping Xerox advance their strategic moves.

PARC researchers also admitted to being incredibly inept advocates of their position during this critical time in the early to mid 1970s. The elitism that Pake and Taylor had encouraged among these research stars tended to get them in trouble with other parts of the organization, as they perceived themselves to be above the corporate fray, money worries included. Worse, they made no secret of their contempt for "corporate types" and "suits" sent to observe their progress. Years later, several of these scientists admitted that their abrasive style made them no friends among the corporate staff—the very people who could have been important champions of their innovations to the rest of the organization. The result was a general "us versus them" mentality that operated throughout Xerox. East Coast executives did not know how to relate to barefoot, long-haired scientists, and PARC researchers blandly disregarded or minimized their relationship with the rest of the organization.

In 1972, Xerox reorganized and created an Information Technology Group (ITG) that assumed overall responsibility for new product development, including the activities at PARC. James O'Neill, another former Ford financial executive, was named as head of the business group. For Jack Goldman and the technology cadre, this appointment was an unmitigated disaster. As Smith and Alexander (1988: p. 154) quoted one senior manager within Xerox:

> *"Numbers bewitched O'Neill even more than McCardell. Archie [McCardell] was attentive and flexible. He would at least listen to others, even if their analysis was more qualitative than quantitative. But O'Neill quit thinking with quantitative analysis."*

O'Neill's impact at Xerox, as head of ITG, was immediate. He had long adopted a mentality of tight control and adherence to budgetary and other financial ratio information as his basis for decision making. If the numbers were wrong, no amount of extra information could persuade him. His approach, which placed tight restrictions on product development decisions and spending, was almost precisely the opposite of how most engineers believed a product development organization should be run. The engineering staff in particular were worried about his priorities, noting that an accounting philosophy, which O'Neill certainly typified, was predicated on risk aversion. Product development, however, does not offer the level of guarantee that someone with O'Neill's sensibilities would insist on. Simply put, it is impossible to accurately predict all of the myriad variables that must be considered prior to adopting new products, particularly when they are a radical departure from Xerox core business.

The appointments of McCardell and O'Neill came at a critical time for Xerox, particularly with regard to PARC and its charter. While not intending to, these appointments provoked the start of a major change in the manner in which Xerox intended to continue its new product development. On the surface, no one advocated a departure from McColough's vision of the "office of the future," in which Xerox led the way with integrated office systems. However, two key decision makers in Xerox's future now had financial backgrounds; with their requisite approach to tight control and quantitative decision making, the seeds of future conflict with the engineering personnel had been sown.

Goldman, as head of research, quickly came to loggerheads with O'Neill and his fascination with the bottom line numbers and avoidance of risk. Fundamentally, they saw entirely different approaches for Xerox's future in new product development. Goldman was a gambler by nature and viewed research and development as a high risk-high return proposition. Jim O'Neill took the opposite view, finding it difficult to justify pouring large sums of money into projects with no clear-cut application, target introduction dates, or projected cash flows. Inevitably, their clashing views led to a feud within ITG, characterized by Goldman's aggressive pursuit of pure research and O'Neill's more cautious approach to new product development: A case of strong "R" and weak "D."

Sadly, no one within the entire top management team was in a position to mediate the quarrel, since Xerox had long since shed itself of senior technical managers at corporate headquarters. Headquarters at Xerox had become the dwelling place of a professional management core, whose long suit was financial acumen and strict control processes and whose inevitable weakness was in the area of technology and engineering. The obvious casualty of this clash was Jack Goldman. At one time reporting directly to the CEO, he endured a series of demotions over the next years, effectively removing a strong, if abrasive, advocate for PARC and its products.

DALLAS VS. PARC

With the write-off of SDS, Xerox began looking around for an alternative source for PARC's product development ideas. Since the original mandate for PARC was to supply SDS with viable computer products, its demise threw senior management into a quandary as they searched for a reasonable alternative outlet for PARC's labors. They settled on the Noncopier Office Products Business Group (NOP) as the best alternative. The NOP had originally been set up in the mid 1960s to develop a wider range of products for office use; in effect, to expand Xerox's integrated office concept beyond simply copiers. From that time until 1973, they had only managed to develop a fax machine and were generally considered a failed venture. It was into the NOP niche, coupled with the decline and write-off of SDS, that PARC found itself linked. In McColough and McCardell's minds, the match made perfect sense for establishing a new source for PARC's sophisticated gadgetry.

While the idea of linking PARC with NOP seemed logical, Xerox's next move was less understandable. In searching for a geographic location in which to establish NOP, now headed by Robert Potter, consultants recommended either of two sites: the San Francisco Bay area or Dallas. Goldman, Pake, and almost every engineer within PARC favored the Silicon Valley location with its obvious proximity to PARC's own buildings. After all, they reasoned, if their true mandate was to provide research and new product ideas for NOP, what would make better sense than establishing them within close physical distance to PARC? O'Neill, however, conducted elaborate financial and logistical analyses which deemed Dallas the better choice for its cost of labor, transportation alternatives, and other financial factors. On the basis of his recommendation, McColough gave his approval for the move to Dallas. Once again, financial analysis and cost control proved to be the deciding factors in a decision that was to have monumental implications for the coming years.

From the start, the attempted match between PARC's free spirits and the more conservative culture of Dallas, was a source of repeated tension. The first battle centered around the push to develop a word processor to perform many of the mundane tasks secretaries are assigned. The IBM magnetic tape system was the current state of the art among businesses, and other companies, including Xerox, rushed to get their own versions to the marketplace. There was just one problem: Potter and NOP wanted an electromechanical model that was noninteractive and nonprogrammable. PARC scientists fought against what they considered to be a step backward in the technology march. As some of the foremost computer designers in the world, they realized that the electromechanical technology was a short-term stepping stone to the next level, involving interactivity and computers. By 1973, Thacker's team had already developed a working Alto computer. The PARC team argued that the opportunity was perfect to leapfrog the dated technology that everyone was entering.

As maddening as NOP's approach was to PARC's researchers, Potter was equally frustrated with PARC's attitude toward what he considered to be Xerox's reason for existence: enhancing and maintaining profitability. His first visit to PARC merely reinforced his opinion that the average scientist was operating in a fog, with no regard for, or understanding of, the pressures that corporate headquarters had put on him. PARC, by their own estimation, needed over a year to get Alto ready for introduction. Potter was facing an immediate window of need: He felt he had to hit home runs from the beginning. In his mind, the electromechanical word processor, while not the best technology, offered the benefits of immediate payoffs in a marketplace that understood the technology. With his own minimal knowledge of computers, he simply did not see how the Alto could offer a viable alternative. Following his decision, Xerox introduced the 800 model word processing typewriter in mid 1974.

Although a decent machine, the 800 was by no means state of the art by 1975. In fact, while it was a reasonable seller, given the high hopes for it, its sales volume was disappointing. The next year, however, Xerox compounded their error when they began searching around for a follow-up to the 800. Two obvious alternatives presented themselves: take a risk and push for production of the Alto, thereby offering a quantum leap forward in office machine technology, or take the more conservative route and develop an enhanced word processor based on an upgrade of the 800 model. Despite the recommendations from an internal study group, McCardell and O'Neill went with the financially "safe" alternative and selected the upgraded model 850 electromechanical word processor. Their reasoning followed that estimates put the cost of producing the Alto at more than $10,000 per machine, compared with a cost of less than half that for the 850. Because top management did not understand computers, they could not see the Alto as anything more than a glorified typewriter saddled with huge manufacturing costs. In contrast, the 850 was a safer bet since it offered a number of enhancements over the older 800 at a lower cost. Again, the bottom line won out.

A Company Wearing Blinders

Xerox's competition was not sitting still throughout the the 1970s. The company's big blunder in sitting on the Alto and refusing to recognize the power of computerized word processing opened the door wide for Wang computers. Wang had developed their machine close on the heels of the PARC scientists and unlike Xerox, with its huge bureacracy, turf battles, and risk aversion, rushed it to market. Archie McCardell and Jim O'Neill were right: The initial cost of the Wang word processor was over twice that of the standard electromechanical machine on the market. However, they were stunned by the industry response to Wang's computer. Price, it seemed, was not the great inhibitor that the financially-trained minds at Xerox had assumed.

At the same time, by 1976 Xerox's preeminent position in office copiers was badly eroding. Their low-end copiers were being challenged by Japanese models that boasted far higher reliability at lower prices. Once Savin/Ricoh had developed their liquid toner technology, they were able to offer their machines at 30–40 percent lower prices than comparable Xerox equipment. The picture was not much better for the high speed, high use machines at the top end of the market. IBM and Kodak had been making steady inroads into this market segment for years, with better quality machines capable of producing clearer copies. Customers began deserting Xerox in droves. For any Xerox executive who cared to look, the numbers were daunting. In 1972, Xerox controlled an unbelievable 95 percent of the office copier business. By 1977, that figure was down to 70 percent and falling fast (Smith and Alexander, 1988).

The next move at Xerox had important implications for their willingness to capitalize on the tremendous development going on at PARC. In 1977, Archie McCardell left Xerox to take over the top job at International Harvester. His successor, David Kearns, was a marketer who had been recruited from IBM, a big switch from the Ford-trained number crunchers as typified by McCardell. Unfortunately, Kearns assumed a position in an organization that seemed to operate increasingly by inertia and less by systematic plan. PARC petitioned Kearns to consider marketing the Alto, which, by now, was being used at over 1,500 sites within Xerox. Everyone, it seemed, had an Alto, with its wide range of word processing, graphics, and reporting software. No one, however, was ready to test the open marketplace.

Xerox's efforts during this critical time were instead focused on strategies to recapture their share of the copier market. Product after product was introduced through the mid 1970s with middling results. In adopting an essentially defensive strategy, Xerox consciously chose to remain the conservative, orderly company that they had become, concentrating on tried and true products in comfortable markets. At a time when the entire theme of the "office of the future" was changing, Xerox ignored the signs and worked merely to resolidify their position within office copiers.

Nearly six years after its development, Xerox had still done nothing with the Alto. The nadir came in early 1978 when, following a lengthy series of proposals and sales plans, Kearns also turned down PARC's request to market the Alto. These decisions cannot be understood without a clear view of the politicized nature of Xerox at this point. Research had become the scapegoat for just about every operating division that was convinced they were spending too much on pure research and almost nothing on development. Break up research, these voices argued, and tie their activities directly to the divisions where we can control the output. Research, for its part, adamantly rejected the idea of being broken up, arguing that Xerox's problems with development had less to do with them than with the corporation's own internal controls and red tape. They viewed the pressure to break up research as the equivalent of killing the goose that lays the golden eggs, all in the name of giving Jim O'Neill and other division heads an alibi for their own shoddy development records.

Kearns' decision, ratified by Peter McColough, was depressingly predictable. Although allowing research to continue to operate as a corporate function, Jack Goldman was moved out of his post and given a ceremonial appointment as chief scientist. George Pake took over as head of research at Xerox. Meanwhile, little changed for PARC. It continued to create products that Xerox management was not sure how to exploit.

By 1981, the Alto was doomed and with it, Xerox's opportunity to capture a market that would prove to generate almost unbelievable profits throughout the coming decade. Xerox introduced a work station version of the Alto renamed "Star," to disappointing sales, principally because they now found themselves locked in competition with a number of other personal computing products in the marketplace. One year later, IBM introduced their PC, which, following a development cycle of only thirteen months, was to revolutionize and dominate the personal computing industry for the rest of the decade. For Xerox, when they finally decided to enter the office personal computer market *eight years* after development of the Alto, it was a classic example of too little, too late. Their machine won them no acclaim from companies that found it incapable of running spreadsheets, slower than competitors' products, and served by an inexperienced sales staff.

The irony of the failure of Xerox to push the Alto was based on their own internal pressure, emanating from McColough at the top, to hit another home run following the phenomenal success of the office copier a decade earlier. The Alto offered Xerox the precise opportunity they had been looking for and they failed to recognize its potential, despite the vociferous cries from scientists within their own organization, who saw the limitless future for personal computing.

WHAT WENT WRONG?

The many causes of the failure to exploit the Alto's technology all stem from the stifling internal environment that pervaded Xerox during the 1970s. Turf wars among the various divisions and functions were rampant, and with them, a strong philosophy of "Not Invented Here" (NIH) flourished. "Not Invented Here" refers to the parochial attitudes among many departments and organizations that an innovation is only good if it was developed within their own specific boundaries. Anything from the "outside," whether external to the overall company or simply from another department, is viewed with suspicion and usually ignored. PARC's mandate had been to perform the sorts of pure research that could be exploited by natural outlets within Xerox—initially under the purview of Scientific Data Systems. Unfortunately, from the beginning, the various operating units had trouble understanding how PARC findings fit into their own strategic goals. With no one taking ownership of PARC research, truly novel ideas like the personal computer and laser printing simply languished, or at best, were used internally but never put into commercial production.

Objectively, it could be argued that Xerox was distracted from seizing the opportunities presented by the Alto at precisely the wrong time with litigation difficulties, economic downturns, and poorly conceived acquisitions. All these events served to direct the top management focus away from internal political turmoil and communications from scientists at PARC. Nevertheless, while the timing of many of these activities was unfortunate, it does not excuse Xerox from overlooking the underlying purpose of their creation of PARC in the first place: To establish a world-class research center for the development of technologies aimed at creating the office of the future. The sin of McColough, O'Neill, McCardell, and others was to ignore PARC's reason for existence.

Another problem was the increasingly narrow focus and fascination with bottom line numbers that plagued many of Xerox's top executives during this time. McColough had given PARC the challenge to "hit another home run" following the now ten-year-old success of the office copier (Smith and Alexander, 1988). When the market for word processing began to take off in the late 1970s, the Alto, which was technologically a generation ahead, promised to be just such a home run. When pushed to the wall, however, McColough and his senior staff froze. Mesmerized by the production costs and other bottom line numbers, Xerox opted for the safer bet and released a mediocre product that left the door wide open for Wang to dominate the word processor market.

The supreme irony of the Alto story is that this product was just the sort of technological innovation for which PARC had been set up. Xerox, when confronted with an opportunity to take a quantum leap forward in office product technology, decided to follow a conservative path and thereby lost out on a golden opportunity. Clearly, in the decade between the development of the first commercially successful copier and the creation of Alto, a fundamental cultural shift had taken place at Xerox. They had moved from a leading edge, risk-taking organization to one in which the operating philosophy was "maintenance." In other words, the goals now were to maintain market share, profitability, and status quo. In the face of increasingly frantic calls from PARC to bring the Alto to market, the 1970s showed a company locked in indecision and if not complacency, at least containment. Alto, and the philosophy driving its development, no longer fit into the mold that Xerox had become.

In the final analysis, was the Xerox Alto, to borrow a phrase from Shakespeare, "a tale told of sound and fury, signifying nothing"? Obviously not, as its technical contributions have had important implications in the larger field of personal computers. Certainly, the Alto represents a sobering example of a brilliant technological achievement coupled with totally inadequate development and commercialization. One of our points in this book has been to make it clear that successful projects are those that are well-conceived, adequately supported, and marketed with foresight; in short, that the project management process does not simply end when a product has been finished. If project management were simply about product development, the Alto would be considered a true success story. Unfortunately, technical achievement by itself is not sufficient. Successful projects are also those that withstand the test of the free market in the court of final opinion—the consumers. And in its unwillingness and ultimate inability to bring the Alto to market, Xerox failed this most important test.

References

Business Week. 1975. "Office Automation has Emerged," June 30.

Dessauer, J.H. 1971. *My Years With Xerox, the Billions Nobody Wanted.* Garden City, NY: Doubleday.

Jacobson and Hillkirk. 1986. *Xerox: American Samurai.* New York: MacMillan.

Smith, D.K. and R.C. Alexander. 1988. *Fumbling the Future: How Xerox Invented, Then Ignored, the First Personal Computer.* New York: William Morrow and Co.

Thacker, C. 1986. "Personal Distributed Computing: The Alto and Ethernet Hardware," *ACM Conference History of Personal Workstations,* January.

17

Quintessential Project Failure: The Ford Edsel

WE SUSPECT THAT there are few names which conjure up images of monumental project failure more rapidly than the Ford Edsel. For years, the Edsel has been used in classrooms and boardrooms as a symbol of disaster. The Edsel symbolizes one of the first important fiascoes in American business. This statement is not to suggest that it was the first project failure, but rather, that its very high profile fall has become so ingrained in our psyches that Edsel will forever be synonymous in corporate lore with project management disaster.

Because of the popularity of this "Edsel = disaster" mindset, it is important to separate the myth from the reality of the Ford Edsel. Contrary to popular belief, the Edsel was not the inept failure it has been made out to be. Rather, Ford actually made several positive and appropriate moves in introducing the car. On the other hand, the Edsel case will illustrate other, more fundamental errors in Ford's management assumptions and actions while engaging in this project. It is as a result of these errors that the Ford Edsel was destined for cancellation less than three years following its much ballyhooed introduction.

One theme that becomes apparent in analyzing the Ford Edsel is that it suffered the death of a thousand cuts. No one management act or environmental disturbance by itself doomed the Edsel. Rather, it was the combined effect of several miscues happening during a critical juncture that ultimately caused its failure. This point is important and was illustrated in the Xerox Alto chapter: Many of the determinants of project success occur after the project has been developed. Successful projects are not simply lauded on the basis of technical achievement. The best project development in the world is useless without strong commercial follow-up. And commercial follow-up represented a key shortcoming in the brief but colorful history of the Ford Edsel.

THE EDSEL'S DEVELOPMENT

Since the 1920s, when Ford surrendered its overall market share lead to General Motors, the company had been known primarily for the low-priced cars it mass produced. Its Lincoln division dealt with high-end automobiles, leaving the Ford name well-known for its niche in the less expensive car market. By the mid 1950s, Ford sought to forever change its image as a "low end" car by making a dramatic entry into the medium-price range, thereby competing directly against Pontiac, Oldsmobile, and Buick at General Motors and Dodge and DeSoto at Chrysler. In contrast, the Mercury, Ford's only entry into this car market, held a 20 percent share of the market.

The Edsel was to be Ford's great response to their low market share in medium-priced cars. Not only did the time seem right, but the general U.S. economy appeared poised for exploitation in the medium-price range. Since the end of World War II a decade earlier, and to a lesser extent, the Korean War three years before, the American population had been enjoying a period of unrestrained, joyous expansion. New ideas, product innovations, and larger and larger markets for these goods seemed to have become a permanent fixture of the American economic landscape. The GI Bill had created a burgeoning middle class, full employment guaranteed high disposable income, and the general optimism of the age seemed to foretell bigger and better things ahead.

In the automobile industry, similar opportunities awaited the companies that had the foresight to understand and act on these trends. For example, one study in 1957 had found that every year, over 20 percent of car owners traded up from their low-priced model to a medium-priced automobile. More importantly, there were crucial differences in brand loyalty depending upon the company. In other words, someone owning a lower-priced GM car, such as a Chevrolet, tended to trade up to a medium-priced GM car—an Oldsmobile, Buick, or Pontiac—87 percent of the time. On the other hand, low-end Chrysler car owners (driving Plymouths) traded up to a medium-priced Dodge or DeSoto 47 percent of the time. Most damning of all: Of low-priced Ford owners, only 26 percent stayed with Ford products (the Mercury) during trade-up (Baker, 1957).

Ford clearly saw important shortcomings in this element of its overall strategic marketing plan. The problem was that Ford simply had never offered a viable medium-priced car and so, had tacitly surrendered the field to GM and Chrysler for the last two decades. With the new affluence of the 1950s, this situation could not be allowed to continue indefinitely as it signaled a potential disaster. Car owners continued to use Ford autos as a springboard into GM and Chrysler ownership. It was against this backdrop that the Ford Edsel had originally been conceived.

In his book, *Management Mistakes and Successes* (1994), Robert Hartley breaks down Ford's efforts toward introducing the Edsel into four distinct efforts: marketing research, styling, creation of a separate division, and promotion. As we will see, in spite of seemingly careful groundwork and project development, a number of factors interacted to take the best intentions and thwart them.

Edsel's Marketing Research

Ford began market research long before the Edsel itself had been conceived. They were fully aware that their hesitancy to enter the medium-priced car market had left open an enormous window of opportunity for exploitation by GM and Chrysler. The Edsel was to be aimed at the young couple, moving up the corporate ladder and ready to step into a higher status car. As part of the direct research to test the waters for such a car (nicknamed the "E-car" at Ford because no formal name had been selected), researchers collected over 2,000 alternative names and test marketed them with focus groups and on-street interviews to determine which fit the bill in creating an image that would appeal to the young, professional target market. Perhaps because of the sheer size of the name list, no conclusive alternative name emerged.

The name of Henry Ford's son, Edsel, had been offered as one of the possible choices for the car. Edsel's three sons, all active in Ford management, uniformly opposed the idea of naming the new car after their father. Further, test marketing and free association with the name Edsel led to such negative reactions as "hard sell," "egg shell," and "pretzel." When the name Edsel was finally chosen from the final ten candidates by an internal committee at Ford, it was not even on the final list, but emerged through an inability to arrive at consensus on any of the other choices (Hartley, 1994).

Styling

Imagine that you were an automobile designer specializing in styling and you were approached by Ford executives with the mandate to give a look to the Edsel that was both distinctive *and* discreet. That was the challenge facing Ford's stylists. They wanted to make a bold statement, separating Edsel from the other competitors, but at the same time, stay within the bounds of taste and appeal.

Styling work began in 1954 and from the beginning was unconventional in its approach. Over 800 stylists were brought into the project at one time or another. Hundreds of sketches were made, altered, modified, and rejected before the final version won the approval of this small army of automobile stylists. One of the well-known features of the Edsel was the famous vertical grill which was intended to remind buyers of the luxury cars of an older generation—the Packards, Pierce Arrows, and LaSalles. It gave the Edsel a unique front-end silhouette that was instantly recognizable. Among the other distinctive features of the car were its large tail fins, lavish uses of chrome and glass, and use of push button technology. The automatic transmission responded to push buttons, the hood and rear trunk latches were triggered by push buttons, and the parking brake was engaged with a push button. The overall effect was intended to be one of high technology and ease of use.

Ford also decided that the Edsel should have an equally powerful engine to go with its distinctive styling. A large, V-8 engine came standard with the car, capable of delivering 345-horsepower. Taken altogether, the Edsel's styling and power train were meant to make a statement in hopes of attracting the younger, up-and-coming customer.

THE EDSEL AUTOMOBILE DIVISION

Ford made another important decision in an effort to separate the Edsel from the rest of the Ford product line. Similar to its separate divisional structure for Lincoln, Continental, Mercury, and Ford, they decided to create a completely independent Edsel division. Their reasoning seemed to make sense: If they truly intended to offer the Edsel not just as an automobile, but as the first in a series of medium-priced cars, they needed to create a complete persona around the car. The result was a complicated search for a dealer network that would be willing to offer the Edsel exclusively. In the end, they succeeded in adding about 1,200 independent dealers, the vast majority of whom would be selling only Edsels on the showroom floors.

Ford's approach to Edsel distribution was extremely comprehensive and left nothing to chance. A total of 4,600 possible dealers had been approached before the final set was narrowed down. Further, decisions on where to place the dealerships were made with regard to demographics, metropolitan characteristics, transportation and other logistics, and dealer reputations. In the end, Ford felt confident that they had laid the groundwork for a smooth and successful introduction of their new car.

PROMOTING THE EDSEL: THE SKY IS THE LIMIT

Ford was determined to make the introduction of the Edsel an "event." In order to adequately fund it, the company allocated $50 million for initial advertising and promotion, a huge amount of money for a new car release. Advertising for the Edsel began on July 22, 1957 with a two-page advertisement in *Life* magazine. These ads did not show the car itself, but were meant to tease the public and gain their curiosity by showing a blurred automobile form racing down the highway. This approach was not a fluke: From the beginning, Ford executives were determined to build the anticipation for the Edsel by keeping it (literally) under wraps. Cars were covered during shipment to the dealers and no prerelease publicity shots of the Edsel were allowed. In fact, it was not until late August that actual pictures of the car were released in advertising to the general public.

In keeping with the stand-alone, nontraditional image they wanted to create for the Edsel, Ford even hired a different advertising firm to handle promotion (Hartley, 1988). This firm was told to present the Edsel as something other than simply "new," as that adjective was considered too commonplace. The entire focus of the advertising campaign was to draw attention to the car, avoid the hard sell, and let the Edsel sell itself (Deutsch, 1976).

Ford made one more decision that was to have serious implications for the Edsel: They decided to jump the usual, late fall starting time for new car introduction (usually around November) and bring the Edsel out early. They wanted to make sure there was no competition from GM or Chrysler with which they would have to share center stage. The fall of 1957 was going to belong to the Edsel exclusively.

THE EDSEL ARRIVES

The actual introduction of the Edsel on September 4, 1957 turned out to be one of the great "non-events" in U.S. corporate history. Following the build-up to its release, the public had eagerly anticipated something quite extraordinary, a revolutionary leap forward in automobile technology. The public's initial response, however, was very "underwhelming." Ford had anticipated orders for the first year to total at least 200,000 cars. This sales volume would have required weekly sales to run in excess of 4,000 cars. The first day showed promise as orders were taken for 6,500 Edsels. However, after the initial flush, sales orders dropped sharply. For the first ten days in October, less than one month after introduction, the Edsel posted sales volume of only 2,750 cars, or about two-thirds of their hoped-for level.

Clearly, something was wrong with the message, so Ford decided to turn up the heat on the car. On a Sunday night in mid–October, Ford spent almost half a million dollars to sponsor a gala television special for the Edsel, signing up such luminaries as Bing Crosby and Frank Sinatra. Although the special did quite well with television viewers, it had minimal effect on Edsel sales, which remained depressed. Worse, momentum was turning against the car as word of its low sales began to filter out. The obvious conclusion drawn by the public was that there must be something wrong with the car to keep it from selling. As a result, bad news continued to breed more bad news.

The year 1958 was disastrous for Edsel sales. Hoping to sell a quarter million cars, Ford ended up selling just 34,481 for the entire year. Even when they introduced a new version of the Edsel in late 1957, with a shorter body design and cheaper price, new sales barely responded.

By 1959, less than two years after its introduction, Edsel's division was formally closed and the car was merged with the Lincoln and Mercury lines into a Lincoln-Mercury-Edsel division. It was a cost-cutting move that was designed to pare down fixed costs until Ford could figure out what to do with the car. In the fall of that year, the third (and ultimately final) models of Edsels were introduced with far less fanfare than had occurred two years previously. Sales were limp and production of the Edsel was officially discontinued on November 19, 1959.

In its two-year life, the Edsel had managed to generate sales of only 109,466 cars, a sad total when compared to initial projections of more than six times that level. In closing their books on the Edsel, Ford took a $200 million loss, representing its initial investment plus advertising and operating losses.

WHY DID EDSEL FAIL?

On the surface, the Edsel should have succeeded. Within the scope of their conscious actions, it could logically be argued that Ford did everything possible to ensure the success of the car. They had spent a long period planning its introduction, they had employed hundreds of people to make sure the designs were cutting edge, and they had approached promotion and distribution with care and creativity. In short, from a project management perspective, the Edsel *should* have been a success. What went wrong?

In critically evaluating the Edsel, it becomes clear that no one mistake torpedoed the car. Rather, as we initially stated, the Edsel truly suffered the death of a thousand cuts—it was the victim of numerous, interacting factors, not one central error. Further, while many of these errors are directly attributable to Ford's actions, it is important to keep in mind that some of the causes of Edsel's failure were beyond their control, resulting in something akin to bad luck.

First, the Edsel's introduction coincided with the first economic downturn in the U.S. in over a decade. Following the collapse of the stock market in late 1957, the 1958 recession was a powerful drag on the economy in general and automobile sales in particular (Reynolds, 1967). Consider the following sales numbers for that period (*Ward's Automotive Yearbook*, 1973):

Year	Sales Volume (Units Sold)
1955	7,920,186
1956	5,816,109
1957	6,113,344
1958	4,257,812
1959	5,591,243
1960	6,674,796

For the year 1958, the entire automobile industry's sales volume was less than 70 percent of the previous year's total. In fact, it was not until 1960 that sales volume figures returned to prerecession levels. Edsel was a victim of this poor timing. One car dealer in a *Business Week* article (1957) noted that medium-priced car sales are typically the most affected by economic downturns. In fact, for the year, automobile sales ended up the lowest since the return of normal post-war industry output in 1948.

Further, as Hartley (1988) notes, the trend in U.S. auto purchases was changing. With the impetus of the recession serving as a catalyst, the average American consumer began looking for down-sized, more economical cars. For example, while 1958 was a dreadful sales year for the Edsel, it was the start of a boom period for Volkswagen's fuel efficient and distinctive "Beetle." Import car sales in general had more than quadrupled since 1956 to over 430,000 units.

Another factor that hurt the Edsel was its image as a powerful, sleek car at a time when the National Safety Council was pushing automobile manufacturers to downplay speed and horsepower in their advertising. In 1957, the Automobile Manufacturing Association, of which Ford was a prominent member, signed an agreement that specifically stated that advertising power and performance would be proscribed. The effect was to eliminate the possibility of promoting two of the outstanding features of the Edsel. The Edsel was intended to conjure images of power and handling at a time when the industry was backing away from these very messages.

Ford was also a victim of its own hype. In its efforts to offer a car that they considered not just new, but revolutionary, Ford had inflated consumer expectations to a point that could not possibly be satisfied. This point became crystal clear when potential customers saw the car for the first time and noted only marginal innovations and improvements. The body, though distinctive, offered no "revolutionary" features, prominent tail fins and push buttons notwithstanding. The engine was large, but not excessively so. The car was well-appointed, but by no means luxurious for the price tag. In short, the car had been promoted to a new level and customers instead saw more of the same old thing they were used to from GM and Chrysler.

Another difficulty emerged with the quality of the cars themselves. In an effort to steal a march on other new car introductions, the Edsel was rushed to the marketplace in September of 1957. This move had two unforeseen effects: First, the car itself was not ready for early release. Several owners complained of oil leaks, rattles, and bad brakes. In fact, the dreaded "lemon" label was being affixed to these cars because of the quality problems that could have been corrected easily enough if more thorough quality checks had been performed.

The second implication of the early release was that timing, initially thought to work for Ford, actually worked against them. Dealers all around the country typically view the period of early fall as a time to unload the previous year's models in order to clear lot and showroom space for the upcoming model year. As a result, Ford found itself competing with 1957 car models that dealers were strongly motivated to move at lower prices. Potential customers, mindful of the stock market problems and looming recession, made their purchase decisions on the basis of monetary incentive, further depressing Edsel's sales.

Another important factor that contributed to Edsel's failure was the separate divisional structure Ford created for the car. Ambitious and optimistic of Edsel's "sure" success, Ford decided not to shoehorn an Edsel organization into the preexisting corporate structure. The Edsel's entry into medium-priced cars was intended to spawn several alternative makes and styles, necessitating an entirely different operating structure to support its growth. Unfortunately, Ford jumped the gun. In creating a separate organization to support a new, untried venture, Ford added millions in fixed and overhead costs to the bottom line. The results were to raise the break-even revenue requirements to a level that was extremely difficult to satisfy, particularly for a new car. As a result, the Edsel needed to not only be successful, but a major hit if Ford were to recoup both its initial investment and ongoing operating costs. Under this scenario, the Edsel could not get by with lagging early sales. There was pressure from the first day to make a major impact and when that did not happen, Ford executives pulled the plug to stop the financial hemorrhaging that Edsel's mediocre sales had caused.

Finally, although Ford's marketing research was extensive, it ultimately led to some profound mistakes. Ford had begun fully researching the market ten years prior to Edsel's development and introduction. As a result, some of the assumptions that guided initial decisions proved to be no longer valid by the time the car reached the marketplace. Some of the styling decisions were derided as passé (e.g., the huge tail fins) while the push for a large, powerful engine flew in the face of a four-year-old trend toward smaller, economical cars such as those made in Europe. Another error grew from the choice of names—Edsel. In spite of the fact that this name did not make the final set of ten alternative choices and it generated negative images in the eyes of the average American, it was selected for this most important project. Since then, Edsel has maintained its negative associative images, but for entirely different reasons.

LESSONS FROM THE EDSEL

What are the project management lessons to be derived from the Edsel fiasco? First, it should be clear that from a project development perspective, Ford generally made the right moves: Their initial project planning was comprehensive, their engineering was state of the art, and they brought the car into manufactured status in a timely fashion. If the actual design and engineering were top-notch, how did the Edsel devolve into such a poor state? As with the previous case involving the Xerox Alto, the failure of Ford in developing the Edsel came after the actual creative work had been accomplished. In short, the problems that led to the ultimate failure of the Edsel were primarily commercial rather than technical.

The Edsel and Alto cases illustrate a fundamental point about project management: the management process continues through to the customer. A project that is an in-house achievement but derives no benefits from the marketplace, is no help to any organization. The story of the Edsel demonstrates this point clearly. Any project is only as good as it is accepted and used by the customers for whom it was developed. The Edsel represents a clear example of this dictum—the quintessential technical achievement mired in poor delivery. Through poor feasibility studies and dated market research, Ford convinced themselves that they had designed a car to fill a niche that by 1957 was no longer there. Through a series of marketing and management blunders, Ford executives compounded their mistakes by over-inflating expectations, creating an entirely new organization to support an unproven car design, and rushing the car to market too quickly, before final bugs and manufacturing errors could be caught and corrected.

In hindsight, the end result is, perhaps understandable. In fairness to Ford and its view of the world in 1957, however, the company was justified in believing that they had created the perfect car to plug a huge hole in their product portfolio. During a time of exuberant affluence and confidence, the Edsel might have achieved some measure of the high expectations Ford held for it. Unfortunately, its release occurred at the worst possible time—an example of building a car with a 1954 mindset and finding that such a market no longer existed at its time of introduction (Warnock, 1980).

Clearly, technical flaw is only one of the myriad causes of project failure. This case, as with the Xerox Alto, illustrates the "hidden" side of project management, in which project success is measured on the basis of overall performance—not just in the laboratory but in the marketplace as well. Certainly cost and schedule matter, particularly when they become so oppressive as to threaten the viability of an entire project, such as the Eurotunnel. On the other hand, organizations forget the third part of the project success equation at their own risk: The client is the final arbiter of new project success or failure. In the case of the Edsel, the message was sent loud and clear.

It is heartening to consider that Ford must have learned its lessons well from the Edsel fiasco. Within four years of Edsel's demise in 1964, they released the Mustang, which was destined to become one of the all-time best selling automobiles in U.S. history. In fact, more Mustangs were sold in its first four months than in the entire twenty-six month life of the Edsel.

References

Baker, H.G. 1957. "Sales and Marketing Planning of the Edsel," in *Marketing's Role in Scientific Management,* Proceedings of the 39th National Conference of the American Marketing Association, June, pp. 128-129.

Business Week. 1957. "Edsel Gets a Frantic Push," Dec. 7, p. 35.

Deutsch, J.G. 1976. *Selling the People's Cadillac, The Edsel and Corporate Responsibility.* New Haven, CT: Yale University Press.

Hartley, R.J. 1988. *Management Mistakes and Successes, 4th Ed.* New York: Wiley.

Reynolds, W.H. 1967. "The Edsel Ten Years Later," *Business Horizons,* 10, 38-47.

Ward's Automotive Yearbook. 1973. Detroit, MI: Ward's Communications.

Warnock, C.G. 1980. *The Edsel Affair, . . . What Went Wrong: A Narrative.* Paradise Valley, AZ: Pro West.

18

Falling Comet: The DeHavilland Story

IN EARLY MAY 1953, Calcutta, India braced for the onset of the summer monsoon season, a time of torrential downpours and breaks in the oppressive humidity and heat. Although monsoon was still some weeks off, tropical storms had been occurring in the Calcutta region for days. With heavy rains in the area, a brand new DeHavilland Comet, operated by the British Overseas Airways Corporation (BOAC) flew off into the afternoon sky. Introduced in 1952, the DeHavilland Comet was hailed as the most modern, most efficient, and safest aircraft to date. Its design was ultraconservative and involved considerable over-design as the first jet-powered commercial airplane. The airplane had a fully pressurized cabin and a very well-designed structure. It was, in effect, a shining example of modern technology. Both DeHavilland and the commercial airlines were excited about the potential that the Comet offered for fast, safe travel.

Six minutes later and only twenty-two miles from Calcutta's

Dum Dum Airport, the aircraft exploded and plunged to earth, killing all forty-three passengers and crewmen on board. There had been no indication of problems and no warning of technical difficulties from the pilots (*Newsweek,* 1953). Investigators from Great Britain and India tended to believe the crash came about due to pilot error coupled with weather conditions. Evidence from the wreckage, including the tail section, seemed to indicate that the aircraft had been struck by something heavy, but without any additional information forthcoming, both the authorities and DeHavilland engineers laid the blame to external causes. After all, it was just the first fatal accident for the commercial jet in one year of flying.

January 10, 1954 was a mild, clear day in Rome as passengers boarded their BOAC aircraft for the final leg of their flight from Singapore to London. Most of the thirty-five passengers and crewmen were probably excited at the prospect of coming to the end of a long flight and looked forward to their landing in London in a few short hours. Tragically, we will never know the passengers' frame of mind, for half an hour after takeoff, when the airplane had reached its cruising altitude and speed, it exploded over the Mediterranean Sea, near the island of Elba. Most of the airplane was lost at the bottom, but amid the flotsam, fifteen bodies were recovered. A local physician, examining the remains, noted (*Time,* 1954a: p. 36): "They showed no look of terror. Death must have come without warning." As a safety precaution, BOAC instituted a ban on the use of Comets until the airplanes had been thoroughly checked over. Technicians could find nothing wrong with the sleek new aircraft, and following recertification, the airplanes were again brought back into service.

Alas, too soon. On the eighth day of April, only sixteen days since the Comet had been reintroduced into service, a third aircraft, operated by South African Airways, was scheduled to depart from Rome's Ciampino airport for the final leg of its regular flight from London to Johannesburg. Although they had the specter of two fatal crashes in less than one year to consider, the twenty-one passengers were upbeat, convinced that DeHavilland had put the aircraft's problems behind them (*Time,* 1954b). Under perfect flying weather, the airplane rapidly gained its cruising altitude of 26,000 feet and its airspeed of almost 400 miles an hour. Suddenly, the flight radio went silent and failed to answer repeated calls. A search of the ocean off the island of Stromboli, Italy turned up an oil slick and some debris. Because of the depth of the water and the time necessary to arrive at the crash site, there was little to be found by search crews. Five bodies were all that were recovered this time, though with an eerie similarity to the victims of the second disaster: Facial expressions showed no fear, as death had come upon them suddenly.

What Went Wrong?

Investigators offered various opinions as to the cause of the accident, including sabotage, pilot error, engine failure, fuel tank explosion, and freak weather effects, such as collision with the jet stream. However, without examining any substantial pieces of the airplane, it was impossible to do more than speculate on the causes. Unfortunately, with three separate crashes of the Comet all occurring within less than one year, continuing to simply write these disasters off as pilot error did not wash. Investigators swarmed over the recovered wreckage of the aircraft and reexamined the pieces of the Calcutta accident. They also conducted underwater searches at the sight of the second crash near the island of Elba. Guided by underwater cameras, investigators were able to collect sufficient aircraft fragments (in fact, they finally recovered nearly 70 percent of the airframe) to make some startling discoveries. Perhaps the most important finding was the recovery of the entire, intact tail section. With this discovery it became clear that the fuselage of the aircraft had exploded. Second, it appeared clear that engine failure was not the cause of the accidents. Engines recovered at crash sites showed no evidence of failure or breakage other than that which would accompany their fall to earth. Another finding was equally important: The wings and fuselage showed unmistakable signs of metal fatigue, later shown to be the cause of failure in all three aircrafts. This point was important because it advanced the theory that the problem was one of structural design rather than simple part failure.

Britain's Civil Aviation Board immediately grounded the entire Comet fleet pending extensive reviews and airworthiness certification. For the next five months, the CAB conducted an extensive series of tests to isolate the exact causes of the mysterious crashes. Using an actual Comet that had been supplied by DeHavilland, testing at Britain's Farnborough Royal Aircraft Establishment (RAE) included repeated and rapid pressurization and depressurization of the cabin and simulation of wing flex with hydraulic jacks. An entire Comet was submerged in a 250,000 gallon water tank to simulate cabin pressurization and air loads across the airframe. Meanwhile, the hydraulic jacks attached to the wings pushed and pulled on both of them in order to examine in-flight flex and bounce. Before the CAB was finished, one Comet had been tested literally to destruction; another had its fuel tanks ruptured. Over seventy complete test flights were made in a third test, and between fifty and 100 test models were broken up (*Aviation Week*, 1954b).

In spite of the seeming thoroughness of the testing, some aircraft engineers felt that simulation could never completely isolate airframe problems. They pointed out that such simulation cannot take into account the wide range of stresses that an aircraft experiences as part of its normal use. Vibration from the engines, contraction and expansion of the skin due to temperature changes, applied air and landing loads, and variations in cabin pressure could all serve to prevent the British Board's tests from being conservative enough (Hull, 1954).

Although the aircraft's designers were convinced that the structure would remain sound for 10,000 flight hours before requiring major structural overhauling, simulations showed unmistakable signs of metal fatigue after the equivalent of only 3,000 flight hours. Even when fatigue levels were revised downward to less than 3,000 hours, at which point DeHavilland suggested the airplane be grounded for thorough maintenance, experts argued that Comets would not be safe beyond 1,000 flying hours, a ludicrously low figure in terms of the amount of use a commercial airliner is expected to receive. Nevertheless, testing of the fuselage offered disturbing indications of the cause of failure. Specifically, cracks began developing in the corners of the cabin windows. These cracks were exacerbated by repeated pressurization and depressurization of the cabin. The investigators noted that this result was most pronounced along the rivet lines near the fuselage windows.

Additional flaws in the design rapidly came to light through the extensive testing. First, investigators cited the possibility of severe wing damage through too-rapid fueling. When the blowoff valve was rendered inoperative, testing found that fueling the airplane too quickly could rip wing ribs loose and cut the wing strength by as much as 20 percent. The blowoff valve, designed to prevent the adverse effects of rapid fueling, was subject to seizing up from ice accumulation, a common phenomenon when flying at high altitudes.

Testing demonstrated that the wings had a low resistance to fatigue. At a number of stages in the water tank and wing flex tests, serious cracks appeared, starting at the rivet holes near the wheel wells, and resulting in rivet heads in the top wing surface actually shearing off. Finally, using radioactive tracers, engineers found that fuel from the tank venting system migrated along the wing flap to within two or three inches of the jet exhaust shrouds, where temperatures routinely operated at 350 degrees centigrade. In spite of these potentially dangerous findings, researchers were convinced that none of these issues played a role in the fatal crashes of the Comet. Instead, they focused their attention almost exclusively on the most likely cause: rapid cabin depressurization due to seal failure.

Engineers and investigators began to find incontrovertible evidence in the pieces of recovered wreckage to suggest that the cause of the sudden disintegration of the aircraft could only have been due to cabin pressure blowout. Engineers suspected that the critical failure of the aircraft occurred following sudden depressurization, when one or more windows were literally blown out of the aircraft. This led to a sudden gyroscopic moment as the aircraft nosed down and began its plunge to earth. Engine failures occurred after the Comet's fuselage failure and were not a contributing cause of the disaster (Hull, 1954).

Although at the time no one would admit it, the handwriting was on the wall. Following several years of design and development and an outlay of several million pounds, the DeHavilland Corporation had successfully developed the first commercial jet airliner. After two years, in which Comets carried over 55,000 passengers over seven million air miles, the Comet 1 was never to fly again.

Comet 1 Mishaps—Keeping Score

- October 26, 1952—BOAC Comet failed to become airborne at Rome's Ciampino airport. Plane destroyed, no deaths.
- March 2, 1953—Canadian Pacific Airlines Comet 1A crashed at Karachi, India on delivery flight. Eleven crew and technicians killed.
- May 2, 1953—BOAC Comet crashed after takeoff from Calcutta. Forty-three persons killed.
- June 25, 1953—Union Acromaritime de Transport (French UAT) Comet landed too far down runway at Dakar, French West Africa. Plane ploughed into concrete abutment. Plane destroyed, no deaths.
- July 25, 1953—BOAC Comet skidded off runway while taxiing for takeoff at Calcutta. Plane's port wing spar damaged.
- January 10, 1954—BOAC Comet crashed off Elba Island after take-off from Rome. Thirty-five persons killed. Parts of plane salvaged.
- April 8, 1954—BOAC Comet crashed off Stromboli after take-off from Rome. Twenty-one persons killed. Wreckage sank into Mediterranean.

(*Aviation Week*, 1954)

The Development of the Comet

DeHavilland had long been respected in the airframe manufacturing industry for its innovative and high performance designs. Coming off their successful work during World War II, in which DeHavilland-designed aircraft had distinguished themselves in their dependability, speed, and overall performance, the company believed that they stood poised on the brink of peacetime success in the commercial airframe industry. Not inclined to simply copy the technology of existing aircraft, which still operated propeller-driven airplanes, the DeHavilland designers and executives accurately perceived that the next generatin of airplanes would be jet-powered. Consequently, they decreed that their newest commercial airframe, tentatively called the Comet, would employ jet power and other leading-edge technology, including a pressurized cabin.

The move to jet propulsion was a logical step. Jet aircraft had been around since the war, when German engineers perfected the designs which resulted in the success of the Me-262, the world's first jet fighter. By the late 1940s, the technology was sufficiently diffused to the world's aeronautical engineers, so it seemed the perfect time to take advantage of it with the newest generation of aircraft.

Jets offered a number of advantages over the existing, propeller-driven airplanes, the most obvious of which was speed. Jets could cruise at nearly 450 miles per hour compared with the 250–300 miles per hour a propeller could generate. For overseas flight in particular, this advantage was important. It could reduce the length of long flights from a mind-numbing two to three days to mere hours, encouraging more and more businesspeople and tourists to use airplanes as their primary method for travel. Further, jets tended to be quieter than propeller-driven aircraft, giving a more comfortable interior sound level and ride to passengers.

DeHavilland engineers sought to create a streamlined airplane that could simultaneously carry up to fifty passengers in comfort, while maintaining aerodynamics and high speed. After working with a number of design alternatives, the Comet began to take shape. Its design was indeed distinctive: The four jet engines were embedded in pairs in the wing roots, at the point where they joined the fuselage. From the front, the aircraft looked as though its wings were literally held in place by the engines. The result was an aircraft of remarkable stability in flight, sleek in appearance, and very fast. Though innovative, this design drew immediate fire from American engineers working on their own models. American designers had long been uncomfortable with the idea of embedding an engine into the wing roots, preferring instead to hang engines from pods suspended beneath wings. They argued that DeHavilland's configuration put the airplane at greater risk from engine failure (such as turbine blades breaking or engine fires). Because of the engine's proximity to both the fuselage and wings, any serious problem with an engine would not only affect airspeed, but could seriously compromise airframe integrity, and consequently, will almost certainly affect the survivability of the aircraft.

Another distinctive feature of the aircraft was the pressurized cabin, intended to maintain passenger comfort at cruising altitudes of up to 30,000 feet. In its original testing for safety, DeHavilland engineers had pressurized the airframe to over five times the recommended air density to ensure that there was a clean seal. Consequently, they were confident that the pressurization system would perform well at its lower, standardized settings. Finally, in an effort to add some flair to the design, each window in the passenger cabin was square, rather than the small, round or oval shapes so commonly used.

DeHavilland's goal was to get their new aircraft to market as quickly as possible, in order to establish the standard for the commercial airline industry with new, jet-powered travel. At first, it appeared that they had succeeded: BOAC ordered several Comets, as did Air France and the military. DeHavilland had also received some interested queries from American airlines, notably Pan American Airlines. Right up until the first crash, it looked as though DeHavilland's strategy was working; they were first to market with a radical new design, using a number of state of the art technologies. BOAC's first nine Comet 1s entered service with the airline on May 2, 1952. The future looked bright.

POST-MORTUM: LESSONS FOR MANAGERS

How did the DeHavilland disaster happen? On the surface, the engineering done on the aircraft was first-rate, and its design and construction were up to modern standards. Perhaps they were too modern. As later events demonstrated, the Comet would achieve a kind of notoriety in the aviation world as a project that had been rushed to market far too quickly, making use of design characteristics that severely affected its in-flight safety.

The principle fault in the Comet's design was that it radically departed from the current standard in airframe development and was probably rushed to market too quickly, before adequate stress and design configuration testing could be completed. In DeHavilland's desire to be first to market with a commercial jet aircraft, they may have allowed themselves to be lured by the prospect of huge profits that such an innovation would generate. In fact, there has been some argument that DeHavilland engineers shaved corners and developed the Comet while operating at the safety limits of load, weight, and speed (Petroski, 1985). As evidence, in the wake of the Comet 1 disasters, DeHavilland hesitated to bring out the Comet 2, because the additional structural reinforcement necessary to correct the fatigue and cabin pressure failure problems would seriously inhibit the aircraft's ability to carry passengers. The implication is clear: The Comet 1 was designed to be as lightweight as possible (apparently too much so) in order to carry the largest passenger load it could.

Further, the square windows were suspected of contributing to cabin failure. Engineers noted that fuselage fatigue testing had never been done in a comprehensive way on airframes employing square windows. If, as investigators suspected, the point of cabin pressure failure occurred at the window seals, it is highly possible that the shape of the window itself was a contributing factor. This point was later born out in final hearings on the causes of the crashes of Comet 1s. Britain's Royal Aircraft Establishment laid the blame for these disasters squarely on the result of metal fatigue occurring at fuselage points by the windows, leading to explosive decompression and immediate catastrophic failure.

The DeHavilland Comet represented an unsafe product in both design and engineering. The fact that a new generation of aircraft carried with it some design flaws is not surprising. Advancing the state of the field in any engineering discipline often involves a large element of risk and potential for failure (Petroski, 1985). Unfortunately, when the stakes inherent in the product are human lives, it is impossible to dismiss the fact that failure happens. Certainly, in the cases of a new stereo component or hair dryer, we can reasonably expect radical new designs to uncover new and often unforeseen consequences. However, airframes, automobiles, and other products that can seriously affect human lives do not afford engineers and companies the same degree of flexibility. We simply cannot afford to learn from our mistakes: The stakes are too high.

In the Comet court of inquiry, several recommendations were made to ensure safer future airframe design. The board, headed by Lord Cohen, suggested that in the future it would be necessary to: 1) develop an economically satisfactory safe life of the pressure cabin by more study in design and experimentation; 2) guarantee that pressure cabins will be entirely safe by devising methods for design and test programs to that end; 3) determine stress distribution and fatigue life by calculation and test of cabin components and complete cabins; and 4) use existing methods of calculation for stress distribution more widely, and verify or amend these with extensive strain gauge tests (*Aviation Week,* 1955). The implicit message, quite clearly conveyed, was that DeHavilland had failed to engage in sufficient testing of their innovative new design prior to bringing it to market.

The Comet case emphasizes the importance of project risk analysis. Risk analysis is the convention of analyzing any project prior to its development from a variety of risk perspectives. As such, there are a number of risk elements to consider. First, it is important to assess schedule and budget risk. Based on the best available information, will the project come in at budget and schedule levels we can be happy with? What could prevent us from attaining those targets? Second, project managers must analyze stakeholder risk. Who are the clients and stakeholders for this project? Am I aware of their priorities and concerns? Have I attempted to manage their concerns based on a realistic understanding of their positions? Stakeholder risk management is one of the principle causes of failure in the U.S. nuclear construction industry. Third, we must consider technical risk. Do we understand the technology that this project will employ? Is it new or well-established? If new, have we developed adequate testing to support our development? And finally, what is the downside risk associated with project failure? Does failure merely constitute an embarrassment to the company or something far more catastrophic, as in the case of the Comet?

Many times, our ability to correct problems associated with project development is in direct proportion to the problems that come to light through the use of risk analysis. In the case of the Comet, it is clear that project risk analysis was sadly lacking, as the engineering staff at DeHavilland made the assumptions that overdesign could take the place of adequate stress testing after the fact.

In the final analysis, DeHavilland's "sin" was one of both commission and omission. Their "first to the market" mentality virtually guaranteed that any problems with the airframe were bound to have catastrophic side effects. Further, stress testing offered ample evidence that the aircraft had some serious design flaws. In spite of DeHavilland engineers' contention to the contrary, it is clear that insufficient testing of the airframe was performed. It took a governmental research lab less than five months to uncover not one, but five or six separate, serious flaws in the Comet 1, any one of which could have potentially caused the crashes. The Comet differs from that other classic example of poor design, the Tacoma Narrows Bridge, in that the TNB at least followed standard design and erred in its application in a physically challenging location. The Comet, on the other hand, is an example of a project that was not only poorly applied, but fatally flawed in its design. It offers a bitter lesson on the perils of technological leaps, particularly to the families of those who died while DeHavilland learned.

References

Aviation Week. 1954. "Comet gloom," Vol. 60, April 19, pp. 14-15.

Aviation Week. 1954b. "Fatigue blamed in Comet crashes," vol. 61, October 25, pp. 17-18.

Aviation Week. 1955. "Comet verdict upholds RAE findings," vol. 62, February 21, pp. 16-17.

Hull, S. 1954. "Comet findings may upset design concepts," *Aviation Week,* Vol. 61, Nov. 1, pp. 16-18.

Newsweek. 1953. "Fall of a comet," vol. 41, May 11, p. 49.

Petroski, H. 1985. *To Engineer is Human —The Role of Failure in Successful Design.* New York: St. Martins Press.

Time. 1954a. "A column of smoke," vol. 63, Jan. 18, pp. 35-36.

Time. 1954b. "Death of the Comet I," vol. 63, April 19, pp. 31-32.

19

Megabucks to Megaprojects

MACROPROJECTS, USUALLY DEFINED as those with budgets in excess of one billion dollars, are becoming more and more common, both in the developing and developed parts of the world. One obvious reason is their allure: Megaprojects imply, almost by definition, huge concomitant advances in technology or improvements in the human condition. Consequently, as more and more money is being poured into fewer and fewer projects, it is important to understand both the up and downsides of these ventures. We will argue that while some of these projects are undoubtedly important and useful contributors to society, others have been initiated for questionable reasons, and are likely to spin on indefinitely, offering results that are marginally useful at best.

The inherent problem with projects of this size is that they rarely deliver the "bang for the buck" that was intended. With the obvious exceptions of the Manhattan Project, which brought an end to the Second World War, the Apollo Moon Project, and some select examples of large construction work, the track record on the use of massive projects to achieve equally massive results has been largely disappointing. In this chapter, we will briefly examine three recent examples of the mega-project: the Superconducting Supercollider, the NASA space station, and the human genome project. Each of these was kicked off with huge fanfare and promises of important discovery. Currently, one project has been permanently shelved, another has been scaled down due to budget constraints, and the third remains so locked in ethical and theoretical debates that its future looks increasingly uncertain.

THE SUPERCONDUCTING SUPERCOLLIDER

Originally conceived in the 1980s as a device to accelerate particles in high energy physics research, the Superconducting Supercollider (SSC) was a political and technical hot potato from the beginning. The technical challenges associated with the SSC were enormous. Its purpose was to smash huge numbers of protons together at near speed of light. That would require energy levels of 40 trillion electron volts. The SSC was designed to operate at a level that would make it the largest particle accelerator ever constructed, far bigger than its counterpart at Fermi Laboratory. In order to achieve these sorts of energy levels, a set of 10,000 magnets was needed. Each of the magnets, cylindrical in shape (one foot in diameter and fifty-seven feet long), must operate at peak levels if the accelerator is to achieve the necessary energy levels for proton collision. The expected price tag just for the construction of the magnets was estimated at $1.5 billion.

The scientific and technical difficulties were only part of the overall scope of the project. Construction of the SSC would be an undertaking of unique proportions. Scientists determined that the accelerator required a racetrack-shaped form, buried underground for easier use. The overall circumference of the planned SSC required fifty-three miles of tunnel to be bored 165 to 200 feet underground. The initial budget estimate for completing the project was

$5 billion and the estimated schedule would require eight years to finish the construction and technical assemblies.

The SSC's problems began almost immediately after President Reagan's 1988 kickoff. First, the public (including Congress) had little understanding of the purpose of the project. A goal as nebulous as "particle acceleration" for high-energy physics is not one easily embraced by the majority of tax-paying Americans. Further, although the original operating consortium consisted of eighty public and private American research centers and universities, it was hoped that European and Asian scientists would wish to conduct experiments with the SSC. Consequently, the U.S. Government's Department of Energy had hoped to offset some of the cost through other countries. While initially receptive to the idea of participating in the project, these countries were vague as to their levels of contribution and time frame for payment.

Another huge problem was finding a suitable location for the site of the SSC. At its peak, work on the SSC was expected to employ 4,500 workers. Once in full-time operation, the SSC would require a permanent staff of 2,500 employees and an annual operating budget of $270 million (*Science,* 1988). Clearly, it was in the interest of almost every state to lure the SSC to their proposed location. The result was a political nightmare, as the National Research Council appointed a special site review committee to evaluate proposals from forty-three states. After making their judgments based on a series of performance and capability criteria, the committee narrowed their list to eight states. Finally, in late 1988, the contract for the SSC was awarded to Waxahatchie, Texas, on a 16,000 acre tract of land south of Dallas. While Texas was thrilled with the award, the decision meant ruffled feathers for a number of other states, and more importantly, their disappointed congressional representatives.

The mounting Federal budget deficit caused problems almost from the beginning. More and more politicians questioned the decision to allocate money at a time when Congress was looking for ways to cut over $30 billion from the budget. This concern ended up becoming a long-term problem, as the SSC was only allocated $100 million for 1989, less than one third of its initial $348 million funding request. Budget battles would be a constant refrain throughout the SSC's short life.

Work continued slowly on the Waxahatchie site throughout the early 1990s. Meanwhile, European financial support for the project was not forthcoming. The various governments privately suspected that the project would never be completed. Their fears were becoming increasingly justified as the cost of the project continued to rise. By 1993, the original $5 billion estimate had ballooned to $11 billion. Meanwhile, less than 20 percent of the construction had been completed. The process was further slowed when Congress began investigating expenditures and determined that accounting procedures were inadequate. Specifically, the General Accounting Office took the SSC's operating consortium, Universities Research Association (URA) to task for poor management and accounting procedures at a time when the overall project was coming under increasing scrutiny and criticism (*Science,* 1993a). In a last desperate move to save SSC funding, Energy Secretary Hazel O'Leary removed URA as prime contractor for the construction project. There was talk of replacing them with a proven contractor—Martin Marietta and Bechtel were the two leading candidates. By then, however, it was a case of too little, too late. Costs continued to climb and work proceeded at such a snail's pace that when the 1994 Federal budget was put together, funding for the SSC had been removed entirely. The project was dead. The nonrecoverable costs to the U.S. taxpayer from the aborted project have been estimated to be anywhere between one and two billion dollars.

The Superconducting Supercollider is an example of the kind of grandiose "big science" project that garners both publicity and notoriety. Few questioned the government's capability of constructing such a facility. The technology, though leading edge, had been used previously in other research laboratories. Rather, the problem was one of overfascination with science for science's sake. The pro and anti-SSC camps tended to split between proponents of pure research and those (increasingly swaying political support their way) who argued that multibillion dollar research was a luxury we could not afford, particularly in an era of budget cuts and hard choices. The SSC position was further weakened by the activities of the research consortium supervising the project, URA. Their behavior was termed increasingly arrogant by congressional oversight groups who began asking legitimate questions about expenditures and skyrocketing budget requests.

The Superconducting Supercollider is an example of a huge, technically feasible project that could not clearly articulate its purpose and potential benefits. It failed because of the shortsighted philosophy that assumed science knows best and should not be questioned. Of the lessons learned in this story, perhaps the strongest is the need to make conscious, up-front efforts to create and communicate a clear mission statement and use it in order to develop and maintain a coalition of supporters. Given the size of the project at a time of shrinking budgets, it is, in the final analysis, surprising that the SSC's developers and supporters neglected to remember this simple message.

THE NASA SPACE STATION

NASA's space station project shares many of the same properties with the Superconducting Supercollider: A hugely-expensive, government funded project for conducting scientific experiments on an extended basis in the weightlessness of earth orbit. The space station was conceived in the early 1980s, though funding, as with the SSC, did not come through until later in the decade, perhaps in response to the success of the then-Soviet Union's Mir space station. From the start, the Space Station has been one of the "prize plums" in NASA's list of ongoing projects, surviving a series of budget battles with congressional watchdogs as it continued to received funding. The project has been projected to cost a staggering $30 to $40 billion by the time it is completed, to say nothing of the continual maintenance of the station once in orbit.

Increasingly, as with the Superconducting Supercollider, the space station has been held up as an example of big science at its worst, given its seemingly insatiable appetite for funds at a time of budgetary cutbacks. Due to political pork, the project has long enjoyed the support of a majority of senators, even in the face of vociferous congressional attacks. That atmosphere has changed in recent years, as budget imperatives have finally made it abundantly clear to all associated with the Federal budget that a pure science project with such an astronomical price tag cannot be protected.

The ax finally fell on the original design in May of 1993 when President Clinton ordered NASA to set aside plans for the $30 billion version and create an acceptable alternative within the next ninety days, costing half as much. Clinton's charge left NASA scientists scrambling, trying to find a reasonable scaled-down version of Space Station Freedom. The three alternatives that they came up with each have their share of drawbacks, leaving the future of the project uncertain. At the same time, some voices in Congress are asking the logical question, "Why not use the Russian's Mir space station to conduct experiments?" This question is a sensible one given the fact that one version of Mir is currently in orbit while another is scheduled to be launched in 1997 (*Science,* 1993b).

Whatever the final decision, one point is clear: The space station cannot be expected to continue as a viable big science project in the face of falling budgets and shrinking appropriations. As one congressman noted of a failed (barely) attempt to kill the project entirely, the current approach of simply scaling back funding is the equivalent of cutting off the mouse's tail one inch at a time (*Nature,* 1995). The project is underfunded to the point that adequate work cannot be conducted, while at the same time, it is still alive in Congress, much to the chagrin of more and more representatives. Compromise is likely to be the worst of both possible worlds: Work will continue slowly under increasing scrutiny, guaranteeing a continued flight of qualified scientists who no longer want to wait for the ax to finally fall.

Will the space station be saved? Perhaps, but certainly in a drastically scaled back version. On the other hand, maybe the more correct question is Should the space station be saved? Will it deliver the bang for the buck that we seek? The answer to either of these questions is that it is unlikely. Runaway budgets are simply not maintainable under our current political climate. At the same time, more and more objective observers are challenging the thought process that has become so enamored of such massive undertakings. Is there not a more important role for science, in bringing to fruition projects that can have a direct and more immediate impact on society?

Obviously, there is a place for pure science, and under some circumstances, such massive projects are appropriate. Unfortunately, as the space station demonstrates, they often do not bear up well under the microscope of public appraisal. The space station, should it finally be canceled, will have been a monumental failure, not of technology, but due to changing expectations in the general stakeholder population. In a time of budget munificence, there is no question that the space station would have been built. Under the current state, it represents an example of governmental spending run amok, fair or unfair though that label may be.

THE HUMAN GENOME PROJECT

One of the most ambitious undertakings that molecular science, chemistry, and biology has attempted is currently ongoing at a number of research sites around the world. The Human Genome Project began in 1985 as the brainchild of Charles DeLisi, former director of the U.S. Department of Energy's Office of Health and Environmental Research. The goal of the project is, quite simply, to develop a "map," or chart of the complete sequence of the three billion nucleotides in the human genome. The implications of the successful completion of such an undertaking cannot be overestimated. It would, quite simply, unlock the very essence of human genetic coding, giving insight into heredity, disease, intelligence, and so forth. According to the scientists at work on the project, a map of the entire genetic code of the human being is the equivalent to demystifying the human condition.

Scientists argue that the potential benefits from mapping genetic sequences are enormous. They include the ability to detect various nonnormalities or diseases in unborn infants, and make necessary genetic repairs in utero, thereby preventing illness before it starts. Imagine a technology that can take a minuscule tissue sample from a fetus, determine the child's likely intelligence, propensity toward problems like Down's syndrome or blindness, and make repairs to the genetic strands to correct these problems. This, in a nutshell, is the potential offered by the Human Genome Project.

Within the United States, the project has been centered at two sites, Lawrence Berkeley Laboratory in California and Los Alamos in New Mexico. The project has led to a tug-of-war between two Federal agencies. Both the Department of Energy and the National Institute of Health are maneuvering for oversight control. The final project is expected to cost anywhere between $3 and $5 billion and take ten to fifteen years to complete (*Science,* 1988b).

The Superconducting Supercollider and space station projects, are both saddled with credibility problems due to the difficulty proponents have in articulating their mission, on the other hand, the Human Genome Project offers benefits that the public can much more easily understand. Put into simple, human terms, the opportunities presenting themselves seem to be nothing short of miraculous when we consider the possibilities to alter the human condition in a tangible way.

The project is still in its early stages and while the technical progress is continuing, the project itself is coming under increasing fire from ethicists and other scientists who question the ultimate implications of the work. Their argument, which is drawing some powerful support from various interest groups, charges that scientists who engage in genetic mapping are setting themselves up to play "God" in a manner that is eerily similar to the government of Aldous Huxley's *Brave New World.* The ability to alter the genetic code of an unborn child leads to some fundamental ethical questions: Who has the right to say what constitutes "acceptable" ranges of human characteristics? Assuming that it is possible to detect and correct the onset of Down's syndrome, birth defects, or other maladies, does that make it right to intervene in the natural order? These arguments are powerful because they level a serious charge: Does anyone have the right to establish a standard of "normalcy" for other human beings? And finally, what happens if this technology is given to the irresponsible who undertake the development of bizarre human mutations in the name of science or national defense?

The debate over the ethics of genetic engineering has led to a split among those in the pursuit of information on the human genome. Some suggest that it is sufficient to proceed with the mapping work in order to develop genetic charting. Others counter that there is no logical point to creating such a map unless there exists the will to undertake the logical follow-up and apply such knowledge. The result is a continued rift in the scientific community as to what the ultimate goals of the Human Genome Project should be. Is it simply an exercise in discovery or is there an implied level of action that should follow? Until these questions are answered to everyone's satisfaction, the Human Genome Project, while likely to yield fascinating findings, will probably never amount to more than a laboratory exercise.

BIG SCIENCE AND BIG BUDGETS

Is "big science" the answer? Critics contend that it is not, arguing that our fascination with "giganticism" more often than not leads to waste rather than achievement. In fact, one Nobel laureate in physics, Phillip Anderson, contends that despite their popularity, big projects have actually become the worst way to arrive at basic discoveries (Bagley, 1988). His point is powerful: from 1945 to 1984, some hundred or more major science facilities, including accelerators, oceanographic ships, and observatories, are said to have cost in all about $24 billion. Now however, a single project, such as a space station, may cost that much.

Further, the allure of big science may actually send such projects into areas where they are likely to do only marginal good, spending large amounts of money for questionable or incremental gains (*Science,* 1993c). Such major projects are so large that they develop a life of their own, including a well-entrenched bureaucracy and administrative core. Short of a major disaster or clearly compelling reason, it becomes almost impossible to shut them down. Further, size does not offer the comfort of problem-free project development. In fact, many times the reverse is true, as the sheer size and number of reporting levels and administrative checks can make it easy for the simplest items to fall through the cracks and be ignored.

The world will not easily forget the Challenger accident on January 28, 1986, when the sudden death of seven people was witnessed by the whole world in horrific and vivid detail. Subsequent investigation demonstrated that this accident was due primarily to a design fault, compounded by management failure (Stallworthy, 1987). Project management involving hundreds, and perhaps thousands of engineers, is not necessarily safe from disaster: size and numbers are no guarantee.

Compare the tragedy of the Challenger and its multibillion dollar budget with the huge success of Apple in the early 1980s. While not expressly naming the company, Dr. Anderson could use the success achieved by Steven Jobs and a small group of anachronistic computer engineers in developing the Macintosh computer as his example of the power of smaller projects. Rather than working through Apple's existing corporate structure, Job's first action was to move everyone to a building off corporate property, giving them a separate, insurgent mentality.

Large projects need not become anathema in corporations. Clearly, there is a need to engage in some huge projects that can have important benefits for local populations. The Aswan Dam in Egypt, for example, has been an important contributor to their economy ever since its construction. Nevertheless, large projects are fraught with large potential problems. These problems include too high initial expectations, a willingness to continue to pour money into an insatiable appetite, and an unwillingness to make the rational, termination decision once serious problems and schedule slippages occur.

In order for megaprojects to function with as high a degree of efficiency as possible, there are some important points that the supporting organization or country must keep in mind. For example, researchers contend that no megaproject should ever be offered the equivalent of a blank check; that is, an undetermined or floating final budget. Further, no large project should be kicked off without comprehensive feasibility studies, including planning for backup systems and fallbacks in case of problems. A more complete list of the important keys to megaproject success includes (Morris and Hough, 1987):

- Good positive attitude by owner, contractor, and others
- Comprehensive and clearly communicated project definition
- Good planning, clear schedules, and adequate backup
- Full attention to quality assurance and auditing
- Proper management of urgency and scheduling
- Proper recognition of external factors
- Political and community support when required
- Full financial analysis and necessary financial support
- Appropriate project organization, contract strategy/terms
- Effective leadership, teamwork, resources, and communication
- Highly visible, simple, and friendly project controls
- Recognition that projects are built by people—none are perfect

Clearly, the above list is comprehensive and deservedly so. The obvious point for readers is that megaprojects and multibillion dollar investments require a degree of care that is far beyond that employed in other less costly projects. Perhaps another way to read the above list is to ask some fundamental questions, such as: Is it appropriate to consider a one billion dollar investment without having performed comprehensive feasibility studies? Should any country willingly invest billions in taxpayers' money without first securing the necessary political support? When framed in this way, it is easy to see why these points are so fundamental to any project investment decision.

KNOWING WHEN TO CALL A HALT

When we review the megaproject failures that we encounter around the world, we also find that there are projects that were canceled, curtailed, or shut down shortly after they had been built. Was that also failure? Not necessarily, because it takes courage and resolution to terminate a project at the right time. In fact, failure becomes a matter of degree. When it is clear that a project cannot achieve its performance goals, does it make more sense to write off the initial investment or continue to funnel money into the project in the hopes of forcing success? There is a well-known military dictum that one should not reinforce failure. Unfortunately, many organizations operate under a philosophy that prevents them from such direct admissions as project failure, for fear that they will lose their jobs.

When the decision to pull the plug on a project is made, more often than not that decision has been delayed far too long. The result is predictable: substantial unnecessary losses are thereby incurred. We argue that the termination of a project at the right time might well be considered a success rather than a disaster. The project manager needs to recognize when the time is right to call it quits based on the particular circumstances of the case. All too often, a project that is an inevitable loss maker is still completed. There have in fact been a number of studies made in relation to the problem of closing down a project early, in order to cut the losses. Balachandra and Raelin (1980) developed a model with twenty-three independent factors, seeking to distinguish between projects that should be continued and those that should be abandoned. It seems that a quick review of the situation is usually sufficient: detailed analysis is hardly called for, and could even confuse the situation.

Failure can be due to any number of causes. Projects operate in a dynamic environment and in the course of their execution, the project manager must constantly review and assess the situation, deciding on his future course of action. Assuming that the project is not going well, these will include:

- abandoning the project
- continuing with the project
- investing more resources to get the project right
- looking for possible underlying causes of the trouble

Likewise, the underlying causes can include the omission of some important factor from the calculations, an unreasonably low estimate, which is a sure recipe for trouble, or temporary problems such as bad weather or a strike. A larger list of the reasons for shutting down a project would include (Bard, Balachandra, and Kaufman, 1988):

- poor initial planning and market proposals
- a better alternative
- a change in company interest or strategy
- allocated time or cost budget had been exceeded
- funds had run out
- key people leaving the organization
- a personal whim of management
- problems too complex for the resources available

There is no question: Abandoning a project is a very difficult decision. It is necessary to consider the project's salvage value, and the costs of closing the project down. Construction projects usually have very little salvage value and very high shutdown costs. It is perhaps for these reasons that financially questionable construction projects are often pursued far beyond the time when, from a rational point of view, they should be closed down (Northcraft and Wolf, 1984).

In making the decision to cancel a project, it is also important to consider more than simply the bottom line financial incentive. One must also work to understand the effects of the cancellation on worker morale. Cancelling a project will introduce a lot of its own problems, such as the need for the reassignment of people and the completion of adequate documentation of the project. This demands that project closure be carefully planned and carried out in an orderly fashion. From a morale point of view, one of the worst approaches is to use a quick "hatchet job," with the sudden withdrawal of funds and personnel.

Of course, to some extent the decision to cancel a project and pull funding are actions that depend upon accurate timing. If the resources are already some 75 to 80 percent committed, one reaches a point of no return. It is better to complete and so get the benefit of the investment. However, even this decision is subject to the availability of funds. If funds have run out, or one of a number of reasons for termination are given, then there may be no alternative. Remember, however, that termination of a project does not always result in extinction. It is sometimes possible to blend the results of one project within another, particularly if there are valuable technological advances that could benefit the new project coming online.

If project cancellation is sometimes the obvious answer, why then are so many projects allowed to spin on, out of control? Unfortunately, many projects are continued even when they are a dying cause. Lockheed's L1011 Tri-star Jet Program was typical of such a project. It was seen by technical observers to be a losing project from the beginning. It was unlikely to earn a profit and it was no surprise that enormous losses were accumulated on the project over some ten years (estimated in excess of one billion dollars). Critics charge that it was the desire to maintain prestige and save face that allowed this project to continue: it was to be a symbol of Lockheed's expanding image in the aeronautical world.

Another factor contributing to the reluctance of management to abandon a project is social pressure and management's reluctance to expose their mistakes and shortcomings. Individuals tend to become closely associated and identify themselves with a project. It is their "baby," and they will act to defend it despite mounting losses and serious doubts as to the project's viability. They continue to "weather the storm," hanging on in the almost impossible hope that they may succeed after all. Withdrawal from these projects is seen by such managers as a sign of weakness (Staw and Ross, 1987).

Sheer inertia is another factor that impedes the decision to close down a losing project. It seems easier just to avoid rocking the boat. Organizational politics and company strategy will also prevent cancellation of an otherwise doomed project. The Deep Tunnel Project in Chicago, designed to serve as a major boost to the city's sewer system, is one such example. The project cost keeps escalating, while the completion date recedes into the far future. Unfortunately, the benefits of the project cannot be felt until the entire system is completed. In all probability, the project would never have been put in hand, had the true costs been known at the beginning. Yet once started, it seems difficult to abandon—there is the continuing hope that it may eventually pay off.

Nevertheless, cancellation is a realistic option. A company that is bold enough to cancel a project and acts early enough, will indeed cut its losses. Such a decision, however, requires as much, or even more, rigour, analysis, and discipline as the decision to launch a new project. Project cancellation should not be seen as a failure, but rather as a key decision in the company's strategy. Cancellation may certainly be the right decision when one or more of the following conditions are satisfied:

- serious cost and time overruns;
- poor technical performance is expected;
- the technical risk is too high;
- the project is found not to suit the company's strategic planning;
- the original project purpose or customer requirements have changed.

We can find many examples where projects have been completed despite the fact that there has been a substantial cost and time overrun, with all the obvious implications in terms of operational profitability. David Davis (1985; p. 96) quotes a case typical of many plant construction projects:

> *"The new plant has had a 100 percent cost overrun and an eight month completion delay and now, six months after startup, it runs at less than half the capacity planned in the original design. The workforce is disgruntled. After a series of production crises, the customers are increasingly impatient, the original project manager has been fired, and the plant manager is feeling shaky."*

Though we know nothing of the circumstances, we doubt whether either the project manager or the plant manager were entirely devoid of responsibility for what had happened. The project manager should have reappraised the project for viability and recommended abandonment if appropriate. Perhaps he did and his recommendation was turned down. Senior management is often reluctant to take a recommendation to review, reengineer, recast, or cancel—but such a decision may well be the right one. However, the number of unprofitable plants that make it to the operating stage serve as continuing proof that their sponsors often balk at this particular decision.

The key is realistic project appraisal. That idea implies a willingness to surrender our belief that money naturally equates to achievement. In fact, the reverse is often the case: escalating commitment to bad decisions by pouring additional funding into them does not improve the decision, it merely makes it a more expensive flop when the time of reckoning is reached. On the other hand, when a project is given strict guidelines, (including clear objectives, time frame, and budget) and enjoys the necessary support of relevant stakeholders (including government and community approval), while working in a creative, team atmosphere, the chances of megaproject success are dramatically improved and with them, the likelihood that the megaproject will achieve the important contributions for which it was initiated.

References

Bagley, S. 1988. "Debating the merits of big science," *Newsweek,* 18 April, pp. 52–53.

Balachandra, K. and Raelin, J.A. 1980. "How to decide when to abandon a project," *Research Management,* vol. 23, pp. 24–29.

Bard, J.F., Bacahanbra, R. and Kaufman, P.E. 1988. "An interactive approach to R&D project selection and termination," *IEEE Transactions on Engineering Management,* Vol. 35, pp. 139–146.

Davis, D. 1985. "New projects—beware of false economies," *Harvard Business Review,* Vol. 64, March/April, pp. 95–101.

Kharbanda, O.P. and Stallworthy, E.A. 1992. *Lessons From Project Disasters.* Bradford, UK: MCB University Press.

Morris, P.W.G. and Hough, G. 1987. *The Anatomy of Major Projects: The Realities of Managing Projects.* New York: John Wiley and Sons.

Nature. 1995. "Space Station survives—for now," vol. 376, August 3, p. 375.

Northcraft, G.B. and Wolf, G. 1984. "Dollars, sense and sunk costs a life cycle model of resource allocation," *Academy of Management Review,* April 1984, p. 22.

Science. 1988a. "Texas lands the SSC," vol. 242, Nov. 18, p. 1,004.

Science. 1988b. "Genome projects ready to go," vol. 240, April 20, pp. 602-603.

Science. 1993a. "University consortium faulted for management, accounting," vol. 261, July 9, pp. 157–158.

Science. 1993b. "NASA rethinks the space station," vol. 260, May 28, pp. 1228–1231.

Science. 1993c. "Big science enters the clinic," vol. 260, May 7, pp. 744–747.

Stallworthy, E.A. 1987. "For want of a rubber seal—disaster!," *The Chemical Engineer,* No. 435, April, p. 31.

Staw, B.M. and Ross, J. 1987. "Knowing when to pull the plug," *Harvard Business Review,* Vol. 65, March/April, pp. 68- 72.

Part 4

Let's Listen and Learn

20

*The Future of Project Management**

INTRODUCTION

What is the future of project management? We are convinced that the importance of project management for organizational success will expand rather than wane in the years to come. In this chapter we will discuss some of the reasons why we see a proliferation in project management and subsequent increasing demand for project management skills in the years to come. Recent articles in popular business magazines such as *Forbes, Business Week,* and *Fortune* make clear that more and more executives have come to recognize the importance of project management techniques for greater productivity and market response. Further, companies need to be-

* Portions of this chapter were adapted from *Successful Project Managers* by Jeffrey K. Pinto and O.P. Kharbanda, New York: Van Nostrand Reinhold (1995).

come more systematic in their approach to developing project managers who have the tools and abilities to get the job done.

In the larger context of international business, attempts to make projections regarding the future are often met with skepticism, wariness, and even amusement, as in cases when prognosticators are confronted with their guesses years later. Peter Drucker, the well-known management consultant and writer, is particularly leery of organizations that become overly infatuated with the future, arguing that those who forecast and make strategic decisions on the basis of probabilities and present-day trends are generally unlikely to succeed in the long-term (Drucker 1992). Nevertheless, the future does hold some tremendous opportunities in addition to its threats and uncertainty. For project management organizations in particular, there are some encouraging signs that point to the likelihood of a tremendous upsurge in market demand for their services and techniques.

THE AGE OF PROJECT MANAGEMENT

"The Age of Project Management" derives from the title of a recent article by a well-known project management scholar (Cleland 1993). In his paper, Cleland painted a convincing portrait of the state of project management in the international sphere and demonstrated the basis for supposing that project management will increase in importance in years to come. We can list the reasons behind the expectation that project management techniques will gain in popularity: 1) dramatically shortened market windows and product life cycles, 2) rapid development of third world and "closed" economies, 3) increasingly complex and technical products, 4) heightened international competition, and 5) the environment of organizational resource scarcity that has led to downsizing and streamlining operations. Let us consider each of these reasons in turn:

1) Shortened market windows and product life cycles

One piece of evidence of the turbulent changes that are affecting the business world is the degree to which product obsolescence has been advanced. Traditionally, in many industries, product launches could be carefully crafted and planned because companies knew that they had a comfortably wide window of opportunity to develop, test, and market new products. For example, the IBM System 360, which so revolutionized mainframe computing, continued as a viable product for nearly a decade. A more recent example concerns the IBM Personal Computer, the PC. This benchmark 64K RAM microcomputer, launched in the early 1980s, remained the industry standard for almost five years. One has only to contrast those examples to more recent PC product announcements and launches to realize that such lengthy product life cycles, particularly in the computer hardware and software industries, are a thing of the past. Indeed, the technology is moving so rapidly today that year-old PC's quickly become passé.

What these and other examples point to is the impact that market timing has made on new product development. One of the authors recently spent several months working with a large computer manufacturer's project teams. Their strategic planning and new product development departments routinely prepare for new product delivery and system modification introductions that are sometimes two years into the future. For example, it was common to hear hardware engineers and project managers developing time frames for future product delivery that had to occur within a three-month market window. If the project was late, it was useless, because a rival would have exploited the opening and offered a substitute product.

These time-to-market pressures and shortened product life cycles have had a significant effect on more and more organizations as they seek to counter such threats through the use of project teams. Cross-functional teams and project management techniques have had a major impact on a number of companies and their ability to deliver new products within significantly shorter time frames. For example, in 1989, Honda was touted for its "superfast" approach to new product innovation. Using team approaches, Honda had shortened the time frame for new automobile development from five years to three years. While a significant achievement, Honda's team approach was just the beginning of innovative process changes in the automotive industry. By 1994, Chrysler President Robert Lutz was able to announce that the Viper Platform (cross-functional) Team had designed, developed, engineered, and produced a new prototype in eighteen months.

Other examples abound. Motorola's order-to-finished-goods manufacturing cycle for its pagers has been shortened from three weeks to two hours. Hewlett Packard can now produce electronic testing equipment in five days compared to the old four-week processing time. These innovations have come about through a realization that dynamic changes in the business environment are forcing companies to become more aware of the need to move products to market at a faster and faster pace, both in order to take advantage of market windows and to allow for technological changes that render products obsolete at an increasingly rapid pace. Embedded in this realization has been the increased use of project teams and project management techniques to maximize organizational resources and creative processes while giving development teams the freedom from bureaucracy and red-tape that can often strangle innovation.

2) Rapid development of third world and closed economies

One of the astounding by-products of the opening of the former Soviet Union, Eastern Europe, and Asian communist countries has been the explosion in pent-up demand within these societies for all manner of consumer goods and infrastructure development. Wherever one turns in examining the developing sectors of the international economy, it is easy to discover examples of projects either underway or about to be initiated. Vietnam has recently opened its borders to a number of foreign corporations and initiated a massive program for infrastructure and industrial expansion. The People's Republic of China, while nominally the world's largest communist state, has been increasingly eager to encourage consumerism and "pockets" of capitalism within its largest cities. Further, project management groups from major industrial construction firms in the United States, Europe, and Japan are in serious negotiations with the Chinese government for a number of large-scale development projects of every sort.

Eastern Europe also stands ready to take advantage of project management in their drive to modernize industries in the wake of a democratic revolution that has replaced the old, command economies with market-driven capitalism. Old, inefficient factories are being closed and torn down or upgraded to turn out new products. While there are very real limits on funding for revitalization of their industries, the nature of the economies of Eastern Europe are such that capital development is likely to continue, albeit slowly at first, for several decades.

Despite the current recession in international economies, the construction industry continues to boom in Asia, which now accounts for over 60 percent of worldwide construction and continues to grow rapidly. This development has been a positive windfall for Japanese, European, and American contractors (*Asian Review* 1994). An October 26, 1989 *Wall Street Journal* article described the enormous capital expansion program underway in Japan—a program that is expected to grow at a double-digit pace well into the decade of the 1990s. It is likely that project management will continue to play a key role in this expansion, not only within the borders of Japan but within other countries that manufacture and market Japanese products (Cleland 1991).

A recent issue of the *Asian Wall Street Journal* (April 18, 1994) carries a fourteen-page supplement entitled, "Asian Infrastructure—Asia Transforms Itself." The article goes to some length to examine many of the most pressing demands for major infrastructure development and expansion. It is clear that while huge spending on infrastructure in Asia is imperative, the governments of these countries simply cannot pay the enormous cost associated with these megaprojects. Consequently, this situation offers enormous scope for the private sector, both local and foreign companies, to exploit these needs.

To illustrate the nature of the level of development currently being undertaken or conceived in these Asian countries, consider some of the examples summarized in the following table, each representing a megaproject forecast to cost at least one billion dollars.

Country	*Number of Projects*	*Cost range ($ Billions)*	*Largest Project*
China	7	1–77	Dam and hydropower station
Hong Kong	3	1–20	Airport
India	2	1–2	Enron power plant
Indonesia	4	1–2.5	Coal-fired power plant
Malaysia	5	1–5.6	Dam and hydropower station
Philippines	1	1–3	Elevated rail system
Singapore	4	1–5	Power station
South Korea	6	1–20	Superhighway system
Taiwan	5	5–17	Mass rapid transit system
Thailand	4	1.5–4.3	Airport

The above is just a cursory list of some of the more exciting mega-projects that are currently being developed in the Asian world. Perhaps even more intriguing is the list of the largest contractors for these projects. We had suggested that the rapid development of these countries offers some lucrative possibilities for private companies, including European and American heavy construction firms. Among the top ten list of contractors for these projects are: Bechtel Corporation, M.W. Kellog, John Brown/Davy, Fluor Daniel, and Brown and Root. As more companies turn to the developing markets of the Pacific Rim, it is likely that new names will be added to this list.

Several of the projects listed are being rushed through to completion at a rapid pace, as they are desperately needed by the governments of the countries. Two different approaches are used by the project organizations in order to move these projects through the pipeline quickly. One model, referred to as BOT, involves the following steps: 1) build, 2) operate, and 3) transfer, as the contractor completes the facility, brings it to an operating mode, and then transfers ownership to the client. The other variant is called BOOT, referring to the steps: 1) build, 2) own, 3) operate, and 4) transfer, in which the contractor takes initial ownership of the facility or project as part of a licensing agreement before eventually turning it over to the government as it is purchased. The latter, in particular, is an exciting concept that has already been translated into reality in several locations (see, for example, Kharbanda and Stallworthy 1992).

3) Increasingly complex and technical products

Many of the products that are being created today in a variety of industries, from children's toys to automobiles, are becoming more technically complex to develop, manufacture, and use. Technologically-driven innovation presents a tremendous challenge for organizations in the areas of engineering and design, production, and marketing. As a result, many organizations are relying on project teams composed of cross-functional groups to create and move to market these products in as efficient a time frame as possible.

Rework cycles are expensive (Cooper 1994) and often come about through poor internal communication between functional departments, all of whom are expected to cooperate in bringing a new product to market. In many instances, the new product introduction process consists of a series of "cycled loops" from one functional area to another. For example, consider a simplified case in which a new electronics consumer product is slated for introduction. Typically, we would see some sort of causal chain set in motion, in which engineering would first design the product and send their specifications to production. The production department, presented with the design details, may object to certain aspects of the products (perhaps due to manufacturing process limitations that do not allow for all the features engineering originally included in the product) and will then return the product plans to engineering for rework.

Following this "loop," engineering may or may not make enough modifications to satisfy production. Perhaps a couple of iterations of this loop will be needed before production is sufficiently agreeable to the design that will be used in prototype development. At this point marketing is finally brought on board and given the opportunity to comment on the prototype. Depending upon their experience with and knowledge of other competing products in the marketplace, they may offer suggestions that will once again cycle the product back to engineering for additional rework, all the while holding up development and new product launch.

The reader can readily see how unwieldy such a causal chain is when an organization is faced with the pressures of new product introduction. Consequently, many of these companies are scrapping this new product introduction strategy and employing cross-functional teams from the beginning of the process. There is ample supporting evidence to show that allowing all relevant departments immediate access to and ability to influence new product designs will significantly shorten time to market delivery.

4) Heightened international competition

Competition drives innovation. It is only in the face of substitutable products that organizations are compelled to upgrade, alter, or develop new and innovative products of their own. In the past, American manufacturing has had the economic playing field to themselves for a variety of reasons, many associated with the impact of World War II on other industrialized nations and the slow industrial advance of developing countries. During this period, U.S. companies had enormous domestic markets to exploit, leading to a sense of hubris which sowed the seeds for later problems. On the other hand, foreign manufacturing was in its infancy, suffering from the teething problems associated with new start-up companies: poor quality, lack of name recognition, uncertain marketing strategies, and so forth. While truly a golden age for American business, clearly this was not a state that could continue indefinitely, in spite of many companies' beliefs that, in fact, it would.

It was not until the late 1960s that any appreciable inroads were made into "traditional" American markets such as automobiles. However, during the 1970s, the oil shocks and Japanese manufacturing skills combined with lower unit prices, served to seriously threaten many strategic and consumer industries such as steel, computers, electronic data systems (copiers), and electronic consumer goods (television). These economic attacks, while painful, offered a mixed blessing in that they served to finally shake many U.S. companies out of the inertia and sense of complacency into which they had sunk.

In many industries, domestic "counter attacks" have been spurred on by increased use of project management. The pressures placed on American firms forced them to develop innovative processes for survival in a new, international marketplace for which they were underequipped and did not foresee with accuracy. Project management has a long history in certain industries, such as airframe development at Boeing and McDonnell-Douglas. As a result of the end of American corporate lethargy, other companies, many of whom had no experience in project management, were convinced to look upon it as a new and useful tool for competitive advantage. One of the more appealing aspects of our project management teaching and consulting experiences has been the exposure to a number of companies in a diverse array of industries that are experimenting with project management for the first time. Properly trained and schooled in its techniques, these firms stand to reap substantial benefits within the international marketplace through speedier product development and greater efficiency of operations.

5) The environment of organizational resource scarcity

Obviously, organizations have never existed in a truly munificent environment in which resource acquisition presented no challenges or concerns. Companies have always been forced to operate in pursuit of a variety of scarce resources: money, trained personnel, plant and equipment, and so forth. Nevertheless, the uncertain economic conditions of the current decade have led to a new management philosophy—belief in the need to "do more with less." As a result of this new approach, many organizations have begun to downsize and streamline operations in pursuit of cost savings and efficiency.

From an operations perspective, the impact of such corporate downsizing has created increased demands on those who remain to perform as effectively as possible within a resource-scarce atmosphere. In this regard, several of these organizations (e.g., Kodak) are relying to greater degrees on project management to provide the dual benefits of rapid product development and time to market within greater cost controls and budgetary limitations. These companies, some using project management for the first time in a formal manner, have discovered one of the important features of these techniques: their use gives project teams the ability to be both externally effective in getting products to market as well as internally efficient in their use of organizational resources.

Project management is predicated on the ability to use resources carefully. In effect, the techniques are themselves "resource constrained." Consequently, in an atmosphere of efficiency and streamlining operations, project management offers a valuable tool for companies to exploit, as many are currently doing.

The above are some of the more compelling reasons why we are likely to continue to witness an increase of interest in and use of project management in international businesses. As Cleland (1991) noted, the strategic thrust of many businesses and, indeed, many countries, points to a continued drive to improve, upgrade, modernize, and develop their infrastructures, markets, and capital and natural resource bases. In this context, the benefits of project management are substantial and clearly equipped to provide these countries and their business organizations with a powerful tool for effective and efficient operations.

At the same time that we are looking with hope to an expansion in the use of project management, it is perhaps ironic to point out the dangers in creating overly optimistic expectations its use as a business technique. It is true that project management is the management of change, versus traditional functional management, which often solidifies the status quo. As a result, it is ideally situated to serve as a platform upon which many organizations can achieve the degrees of flexibility and efficiency needed for long-term survival and prosperity. Because much is expected from the project management movement, it is also helpful to consider some reasoned words of warning on how to avoid the "flash in the pan" sobriquet so often attached to the latest management technique. As Barnes and Wearne (1993: 138) write:

> *Project management is the management of uncertainty, and its future must itself be uncertain . . . no technique with a distinctive name achieves what its enthusiasts hope for it or lasts as long as they expect. The same could be true of project management itself.*

While we may hope and believe that the authors' concerns are overstated, they nevertheless strike an appropriate note of caution that must be considered. The worst future for project management would be to create a new organizational "buzz word" out of the technique, leading too many companies with too high expectations to begin jumping on the bandwagon in hopes of achieving quick and painless solutions to their problems. Like any other useful management tool or technique, project management will work to the degree that organizations employing it do so in a measured and thoughtful approach, understanding its strengths and limitations. Overly ambitious programs, without the necessary level of commitment and support, could do the worst possible damage to the project management profession, by turning the techniques into simply another in a line of "faddish" techniques that were tried by ill-informed and unprepared companies, failed predictably, and dismissed as ineffective.

CONCLUSIONS

In an area of soaring costs, rapidly increasing complexity, and diminishing natural resources, organizations find that they must make do with less and less. Corporate profitability and long-term survival are predicated on their ability to continue to grow, to offer the public new and innovative products and services, and find competitive niches that enable them to survive and prosper. It is within this context of diminishing resources, cost cutting, streamlining operations, and slashing overhead and excessive personnel, that project management techniques can offer a lucrative form of competitive advantage for those firms that have taken the time to learn and use it well.

Project management, a philosophy and technique based on the

ideas of performing to maximum potential within the constraints of limited resources, offers a logical and attractive method for increasing profitability in a number of areas of business. Further, the rapid expansion of Asia, Eastern Europe, and Latin American economies will continue to drive a concomitant expansion in development, building, infrastructure repair, and improvement, as well as within the industrial sectors of their economies. Many, if not most of these pushes for expansion will be fueled by project management techniques.

With the future so potentially bright for expanding the role of project management on a worldwide basis, the only clouds on the horizon concern the ability of more and more governments and businesses to perform these techniques well. As we have noted throughout this book, the lack of formal training for many future project managers is worrisome and must be corrected, particularly through the development of a common skill set and body of knowledge. The rise of project management as a profession is likely to become a key element in these countries' abilities to use the techniques to their maximum potential. As a result, we have devoted the next chapter of our book to examining the professional project manager. Understanding the nature of creating and maintaining a professional body of project managers is crucial to the continued development and industrial expansion around the world.

References

Asian Review. 1994. Construction industry remains buoyant: 31.

Barnes, N.M.L. and S.H. Wearne. 1993. The future of major project management. *International Journal of Project Management,* 11: 135–142.

Cooper, K.G. 1994. The $2,000 hour: How managers influence project performance through the rework cycle. *Project Management Journal,* XXV (1): 11–24.

Cleland, D.I. 1991. The age of project management. *Project Management Journal,* XXII (1): 19–24.

Drucker, P. 1992. Planning for uncertainty. *Wall Street Journal,* July 22, A12.

21

Conclusions: What Does Failure Teach Us?

WHEN ANALYZING MISTAKES and their principal causes, there are two important lessons that should be apparent to every careful reader. First, all organizations, no matter how successful they have been or will continue to be, make mistakes. Ford Corporation, for example, has had a history of highs and lows in this regard. They dominated the automobile world with their Model T and mass production techniques, then allowed themselves to stagnate to the point where General Motors took over the number one position. They followed the disastrous Edsel four years later with the hugely successful Mustang. That is the nature of business events—the cycle moves through both highs and lows. For every project success, there will always be one or more failures.

The second lesson is equally clear: Where there is failure, there is the potential for learning. Unlike the first lesson, which is obvious to most of us, the second may be threatening. It says, in effect, that failure is not to be pushed aside, but studied. Learn from mistakes—learn how not to do it. W. Edwards Demming's famous dictum on quality is to get the process right and then repeat it. The reverse is also true: Learn what did not work and then avoid it in the future. In either case, make the result, unpleasant though it may be, an opportunity for personal and organizational learning. Sometimes the project failures are so small in scope that their losses can be quickly erased. Other times, the failure is more monumental, resulting in long-term or even permanent pain. In no case, however, should such failures be forgotten.

A Quick Guide to Ruining Your Project

We noted at the beginning of this book that not every project deserving of success achieves it. Conversely, not every project heading for the scrap heap arrives there. Clearly, there are a number of events beyond the control of the project team and parent organization that can help or hinder a project's chances of success. Nevertheless, when we consider those activities and decisions which can play an important role in a project's failure, our research and experience with such well-known cases as those listed in this book point toward some important contributing causes of project failure. We do not claim that this list is totally comprehensive, but it does sketch some of the broader and more important lessons to be learned. Consider the following twelve sure-fire methods for ruining your project's chances of success:

How to Ensure a Project Will Fail

1) Ignore the project environment (including stakeholders)
2) Push a new technology to market too quickly
3) Do not bother building in fallback options
4) When problems occur, throw money at them
5) Let new ideas starve to death through inertia
6) Do not bother conducting feasibility studies

7) Never admit a project is a failure
8) Overmanage your project managers and their teams
9) Never, never conduct post-failure reviews
10) Never bother to understand project trade-offs
11) Allow political expediency and in-fighting to dictate important project decisions.
12) Make sure the project is run by a weak leader

Let us elaborate on each of these:

1) Ignore the project environment (including stakeholders)

One of the best ways to consign a project to almost certain failure is to manage it without regard for the organization's external environment, including those project stakeholders who can play such an important part in a project's success or failure. Project "stakeholders" is the term used to refer to any group, internal or external to the company, that has an active stake in the project's development. As such, stakeholders include clients, the overall marketplace, internal functional departments, top management, the project team, and external groups who have been termed "intervenors" by Cleland (1988). Intervenors include any environmental, social, political, community-activist, or consumer groups that can have an impact on a project's successful development and launch. To ignore the potential power of such stakeholder groups is foolhardy and often results from either ignorance or complacency on the part of the developing organization.

Consider, for example, the case of the Bailly nuclear power plant proposed by the Northern Indiana Public Service Company (NIPSCO) in 1972 (Obermeyer, 1990). It was originally intended to be sited adjacent to the Indiana Dunes National Lakeshore in northwest Indiana. NIPSCO acquired all necessary construction permits and began work on the site, in the belief that it was important to build such plants to shelter its customer base from the rate increases due to an overdependence on oil. At the time, the idea of the plant seemed to make sense, particularly in light of the "oil shocks" that were to become commonplace throughout the remainder of the decade.

Whether or not the idea made sense, neighbors of the proposed new facility had other ideas. Several neighborhood opposition groups formed, originally composed of affluent home owners who found they would be adjacent to the nuclear power plant. Once united, they formed the "Save the Dunes Council" and set about constructing a legal minefield to oppose the licensure and operation of the facility. In time, these original opponents were joined by more and more environmental and special interest groups who filed their own writs in support of the Save the Dunes Council's position. Their concerns ranged from safety (they termed NIPSCO's evacuation plan unworkable) to concern for the environment (their fear was contamination of a large section of Lake Michigan).

At the heart of the dispute was the difference in goals between NIPSCO and the Save the Dunes Council. The public utility had a goal of providing cheap and long-term energy for its client base. The intervenor groups sought to maintain the pristine quality of life along the lakeshore while protecting themselves and their property values from unwelcome intrusions. After years of legal sparring in which the case went all the way up through the Supreme Court (where NIPSCO won), the utility determined that the continual legal and social battles in support of the facility were likely to continue indefinitely. Finally, in 1982, after spending ten years trying to develop the nuclear power plant, NIPSCO withdrew its proposal. All that remained was a very large hole that NIPSCO had excavated at the proposed site, and a total cost to the utility of over $200 million.

This case is simply one more example of the problems that can occur when project organizations forget their client base or assume that they know more than their stakeholder groups. One clear message that comes through time after time is the prevailing power of such stakeholder groups to aid or thwart a project's successful development. The corollary is to bear in mind that not all stakeholders are external to the organization. Many projects have been derailed due to opposition (either overt or covert) from other functional groups or operating divisions. Prior to the "go" decision, one highly important factor must be considered: the receptivity of the organization's internal environment to the proposed project. If there is the faintest suspicion of disharmony, it is important to take time to reassess the reasons and take corrective action, including working with stakeholder groups to understand their concerns or making necessary adjustments to the project.

2) Push a new technology to market too quickly

New technologies imply new and unknown risks. Sometimes the allure of being first to market with a new technology causes companies to cut corners, marginalize safety factors, or make quality trade-offs. In the end, these decisions almost invariably come back to haunt that company's executives, sometimes with tragic results. The Tacoma Narrows Bridge employed a well-understood technology (suspension bridge construction) in a new way with unique physical characteristics (a hugely long but narrow structure set over a natural "wind tunnel" site). The result was to push "well-known" technology beyond the breaking point, resulting in catastrophic failure. Likewise, DeHavilland's desire to be quick to the market with a commercial jet resulted in the Comet, a radical design that ultimately killed scores of people and was withdrawn from the market.

New technologies are very tempting to exploit for exactly that reason: They *are* new. They offer the company a leg-up on competition. Unfortunately, in our rush to push these new designs or technical achievements, there is a strong likelihood of inadequate or cursory pre-testing that can result in disaster. There must be a proper balance asserted between being the first to market a product and ensuring that the product will perform in positive, expected ways. Quality has been defined by Genichi Taguchi, a well-known Japanese engineer and writer, as avoiding "the loss a product causes society after being shipped" (quoted in Evans, 1993: p. 44). Taguchi's message is clear: When projects are developed too rapidly, to the point where there are potential questions regarding its performance, that project poses a threat to society and hence, cannot be considered to be a quality offering.

3) Do not bother building in fallback options

All projects run into trouble at one time or another. The question is not whether problems will occur but rather, to what matter of degree. One of the tests of good project management is how quickly the project is brought back on course. This point is important because it disputes the notion some managers have that "good" project managers are those whose projects never get into trouble. That belief is patently untrue. Not all problems are foreseeable. Consequently, the true test of successful project managers is their flexibility and capacity to respond in clear-headed ways to problems once they occur.

A logical exercise in which project managers must engage is to continually ask a series of "what if?" questions. These sorts of questions force the project manager and team to proactively search out likely problem areas rather than waiting for trouble to find them. An important side note: Research has shown that the project managers who spend adequate up-front time developing a series of "what if?" scenarios and their responses to them, are more successful than those who operate in a purely reactive manner, waiting until problems occur before weighing their various responses to them. In fact, one of the authors once conducted a large-scale study of project failure and found that the number-one cause was the lack of adequate trouble-shooting. The study demonstrated some of the salient points we have discussed in this book; primarily, that problems are inevitable. However, the successful project managers are those who are best able to adapt to the new situation with flexibility, look for opportunities, and bring their projects back up to speed rapidly.

4) When problems occur, throw money at them

Problems with projects, particularly on a large scale, tend to induce an element of panic from top management. This panic usually takes one of two recognizable forms: First, heads begin to roll, starting with the project manager and anyone else who is clearly seen as part of the decision making team. In the absence of obvious incompetence or misbehavior, this reaction should be clearly considered before it is acted upon. Frederick Brooks, in his classic, *The Mythical Man-Month,* demonstrates that personnel changes in the midst of a project, particularly when an element of urgency has crept in, are almost invariably counter-productive. Because of the learning curve, it takes that much longer to bring new personnel up to speed on the project, delaying it further into the future. Prior to making personnel changes, ask the question "What do we hope to accomplish by this change?"

The second common mistake is to overreact in the belief that throwing additional money at the project will somehow "buy" success. While this response is also understandable, it is an action that should be taken only after considerable thought has gone into it. Our experience here has been that unless a project truly is suffering from a dearth of funding, increasing its budget will usually not bring the kinds of returns we hope for. The money will get spent, of course. But the larger question is whether or not the company will receive due value for the additional monetary support. The answer to this question is much more difficult to assess, but generally, our experience has been that the extra money given to a troubled project does not necessarily correlate with an improved likelihood of success.

One of the common threads that runs through many of the project failure cases listed in this book is the unwillingness to back away from a poorly managed development process or product introduction, even when the project manager, team, and top management know the project is in trouble. This phenomenon is often referred to as escalation of commitment to a bad decision. In essence, the theory demonstrates that more often than not, managers do recognize the serious (even fatal) problems that exist in their projects. Nevertheless, there is a strong tendency to follow the prescribed course of action in the face of this failure. Worse, it is common to actually commit more and more resources to a losing hand. Research bears out this point: Managers are usually loath to admit to a bad decision and will actually continue to support that course, even in the face of compelling evidence of failure (Brockner, 1992).

One final point: It is important here to distinguish between adding resources to a project that is in trouble and simply reacting in a "knee-jerk" fashion by increasing funding. It is true that additional resources in the form of personnel, support, and money can help a project if conscientiously applied. This is particularly true in situations in which initial funding was too low, throwing the project's completion into question from the beginning. Prior to simply reacting in a panic-mode to project troubles, the first step is to conduct a realistic analysis of where the project currently is, how it got there, and how additional funding can bring it back on target. The project manager is the one who needs to sell top management on the need for more funding. That "sale" can only lead to reasonable returns if the request for money and how it will be used productively is well thought out.

5) Let new ideas starve to death through inertia

The flip side of pushing new technologies out the door without having spent adequate time assessing problem areas is allowing new products to remain in a holding pattern indefinitely. During the 1970s, Xerox, through a combination of political in-fighting and bureaucratic roadblocks, never developed the Alto personal computer into a commercial product, thereby sacrificing millions in profits over the next two decades. How did a company create a sure-fire winner and then sit on it? Certainly, organizational inertia played a role. There was no obvious avenue for bringing these products to market and Xerox executives lacked the will to take a gamble. Further, the training of their top management team was predominantly financial: The numbers game, with its low propensity for risk, enthralled them. The result was Xerox's apparent willingness to forgo huge profits in order to take the safe route. The irony, of course, is that this same company had made its reputation by taking a large risk in introducing the model 914 copier in the early 1960s, and at the same time, revolutionizing office technology.

In 1979, Steven Jobs, the aggressive, energetic founder of Apple Corporation, was given a tour of Xerox's research facility in Palo Alto and along the way, he was shown a demonstration of the capabilities of the Alto. He was amazed that Xerox had failed to market the machine which he was sure would set a new standard in the personal computer industry. What, do you suppose, would have been his reaction were he told that Xerox had developed the machine almost six years earlier and had been sitting on it ever since? Jobs' sense of disbelief is understandable. Because they lacked the will to take a risk, Xerox lost a dominant position in personal computers. The rest is history.

6) Do not bother conducting feasibility studies

Why waste time checking to determine if a new technology will work? Why worry about harmful side effects? Why bother considering customer concerns? Obviously, the answer to each of these questions is because failing to do so is one of the surest roads to project failure. Feasibility planning implies that an organization has done their up-front homework in order to put themselves in the position to conclude a project successfully. Feasibility studies require that project managers and upper management devote sufficient time to understanding the project's risk analysis, cost analysis, time frame to completion, stakeholder analysis, and other relevant information is approved before funding. Obviously, the real danger of such analyses is that they operate under a "garbage in—garbage out" philosophy; that is, one where someone purposely loads up the evidence in one direction or the other to support an a priori attitude.

Some years ago, a team of mid-level managers at a large U.S. corporation was formed into an internal study group and given an assignment to assess the feasibility of investing in the development of a new product. Following extensive analysis over six months of study, the group reconvened to present their findings. The only observer at the first presentation was a corporate vice president who listened impassively for the first five minutes. Once it became clear that the study group's recommendation would be in favor of funding the new venture, the vice president quickly interrupted. "You came to the wrong conclusion!" he said, dismissing them to "rethink their position" prior to the presentation before the full executive committee the next day. As you might imagine, their presentation the next afternoon strongly recommended against funding the project.

The benefit of accurate and reasonable feasibility planning is that it locks the company into a mode of planning, then execution. This approach is equivalent to the "Ready, Aim, Fire" model that typifies effective companies. The alternative, "Ready, Fire, Aim," leads to an incredible amount of waste as project after project is initiated with only minimal up-front assessment.

7) Never admit a project is a failure

Sooner or later, every project will turn itself around, right? Wrong. Many projects fail through mismanagement, miscalculation, or fundamental changes in the external environment. To continue to push a project through to fruition regardless of whether or not it is still viable, is obstinacy bordering on foolishness. It is equivalent to the well-known story of the optimist who, when placed on a large pile of horse manure, began enthusiastically digging in the belief that there must be a pony down there somewhere!

One of the most difficult lessons to learn about managing projects is when to quit. Making termination decisions is extremely difficult, particularly as it must often be done in the face of resistance from the project manager, team members, and upper management proponents. Their opposition is understandable because they have a personal, ego stake in the project. Consequently, they keep digging, convinced that somewhere under all the detritus of escalating costs, poor performance, and sliding schedules, there must be a pony somewhere.

A willingness to acknowledge an error is an important requirement of the termination decision. Admitting such errors and wiping the slate clean may be financially painful in the short term, but it is, in itself, a form of success. AT&T's recent admission that its $7.5 billion investment in acquiring NCR Corporation five years ago was a mistake will work more to their advantage than discredit in the long run. When similar project problems are apparent and irrecoverable, it is important not to throw good money after bad. When the patient is past revival, acknowledge it, learn the relevant lessons, and move on.

8) Overmanage your project managers and their teams

Large corporations, loaded with layers of oversight and bureaucracy, are increasingly becoming some of the worst settings for achieving cutting-edge innovation. We see the same phenomenon with so-called "big science" projects, involving hundreds of researchers, billion dollar budgets, and bloated bureaucracies and administration. Is it possible to achieve greatness inside a large corporation? Certainly, but the odds are stacked against you. Consider the example of IBM who, for years, regularly devoted 10 percent of its revenues to research and development. In 1989 alone, "Big Blue" spent over $6.8 billion on R&D, almost the equivalent of what was spent in all of Japan for that year. And yet, in spite of these huge expenditures, innovations never seemed to find their way to the marketplace. The sheer size and inertia of the organization made it virtually impossible to react quickly or expedite technology transfer to exploit commercial opportunities.

The term "lean and mean" has come into our vocabulary regarding the types of organizations that enjoy better than average success with new product development. The lean and mean organization is one that has not layered itself in the cloak of bureaucracy; rather, it is one that is flexible, and has pushed decision-making authority down to low levels where project managers can make product development decisions without endless rounds of review and modification. The lean and mean philosophy has the potential to transform the American corporation precisely to the degree that companies practice what they are beginning to preach.

Companies must begin to ask themselves how the internal steps of checks and balances are involved in bringing new products to market. Do they suffer from excessive bureaucracy, poignantly termed "staff infection" by one interviewee? The answer to these questions will go far toward determining the flexibility and reaction time necessary to bring products to market in an opportunistic fashion.

9) Never, never conduct post-failure reviews

What can we possibly learn from a failed project? We have heard this question voiced many times, usually by managers who are frustrated and/or embarrassed with the leftovers from a failed project. The first inclination is to sweep the results under the rug as quietly as possible and then move on as though nothing happened.

Failures are written off as flukes due to events beyond our control. Obviously, as the title of this book implies, we strongly disagree: Mistakes are natural. Learning from them without an occasional push, however, is a trait that is much harder to acquire.

While denial is easier, it is the worst attitude one can possibly take toward business in general and project management in particular. Failure teaches us a number of valuable lessons, provided we can review them objectively and nondefensibly. For example, some of the most effective heads of project management organizations are those individuals who can painstakingly walk their project managers back through the development process of a failure to see where the wheels fell off the cart. The process should not be accusatory, but instructive. Further, one of the best techniques we have ever witnessed was one project director who developed a chronicle of past failures and their causes. He became so attuned to the evidence of potential failure that he was often able to detect potential problems before they had become apparent to others in his organization.

Consider the alternative: Ignore the evidence and lessons of past project failures and treat each situation and challenge as though it were unique and not previously understood. The results are predictable—they point to the difference between a manager with ten years of experience and another with one year of experience ten times. Clearly, such an attitude cannot be in the best interests of the organization nor does it help a project manager's career, particularly in the long run. Learning from mistakes becomes more than simply a personal luxury. From the organization's point of view, it is a duty.

Rita Mae Brown once defined insanity as doing the same old things in the same old way, and this time, expecting a different result. If we continue to refuse to learn from the past, not only are we condemned to repeat it (to paraphrase Santayana), but we perpetuate a cycle of personal and professional failure. In the final analysis, perhaps that is the true insanity.

10) Never bother to understand project trade-offs

Like it or not, when managing projects, we are often faced with a series of unappetizing alternatives. These trade-offs often come down to a "dollar–day" determination. In other words, to what degree are we willing to sacrifice money in exchange for our schedule and vice-versa? This question points to the nature of project trade-off decisions: They are frequently balancing acts among rival (and seemingly equally compelling) demands. Do we understand the implications of crashing a project? Have we taken the time to consider the budget impact of such a decision? If the answer to either or both of these questions is no, clearly we are not making decisions on the basis of rational insight. Hard decisions are the perquisite of project management. Uninformed decisions, however, are its bane.

11) Allow political expediency and in-fighting to dictate important project decisions

It will not surprise most canny readers that many operating decisions are made with less than perfect motives; that is, the desire to maximize corporate success and profitability. Unfortunately, in the politicized environment of most organizations, any number of potentially momentous decisions are motivated far more by personal agendas than any desire to satisfy overall corporate needs. Examples of this phenomenon abound: Following the breakup of their telephone empire in the early 1980s, AT&T determined that they needed to create a new marketing and sales-based mentality. They hired as their corporate vice president a former marketing executive from IBM. AT&T had always been dominated by its R&D function in strategic decision making and they were not about to willingly abrogate their preeminent position in the organization. What followed was a period of active political in-fighting, as R&D sought to subvert any moves by marketing to alter the culture and focus of AT&T's strategic mission. Finally, the inevitable occurred as the ex-IBM executive resigned his increasingly untenable position.

What is often lost in this story is the central point: Objectively, AT&T *did* need to reshape their strategic focus; they were about to go head-to-head with a number of upstart long-line competitors such as Sprint and MCI. Nevertheless, R&D saw any strategic shift as threatening their position, so they actively opposed the change, regardless of the negative effect on the overall organization's profitability. That is the nature of political decision-making—it typically emphasizes parochial needs, even at the expense of overall organizational effectiveness.

Project decisions that are made on the basis of power-plays and maintenance of executive prerogatives are bound to be less than effective. In effect, the project becomes a hostage pawn in a larger, more personal game of acquiring and keeping power. Under such circumstances, it is not surprising that excessively political environments have a much more difficult time in successfully developing innovative projects.

12) Make sure the project is run by a weak leader

Obviously, the term "weak leader" is oxymoronic: Successful leaders exhibit many traits, but a fundamental weakness is not one of them. Leadership is an essential ingredient in project success. To borrow a concept from the physical sciences, projects, if left to their own devices, tend to run toward entropy. That is, the natural project state is more often chaotic and disordered than logical and pragmatic. In the absence of a strong leader to keep the project team operating on track, most projects begin to experience the vacuum of indecision—orders are given and rescinded, and a general sense of aimlessness surrounds them. Weak leaders are not merely unhelpful to a project's successful completion, there are actively counterproductive. In the entropic state into which a project can easily fall, money and time are wasted and productivity is minimized, all because there is no firm hand at the tiller.

The project leader is the one person who has the power to make the project succeed by marshaling resources, motivating team personnel, negotiating with stakeholders, cheerleading the development process, and constantly keeping an eye on the ultimate prize, the successfully completed project. Naturally, when described in these terms, it is no wonder that successful project managers are a special breed, one that needs to be carefully cultivated and guarded within our companies. Their role in successful project development is almost always highly visible. Conversely, in the preponderance of projects that failed, the project manager either was essentially invisible to team members or exhibited the worst sorts of characteristics a project manager can have: weakness and laxity in place of decisiveness and determination.

The above is a list of conclusions that we have drawn from the various cases presented in this book. While no one project failure studied here violated all of the above twelve points, it is easy to see how these examples apply to varying degrees, to just about each case we have analyzed. In the end, what are the conclusions we can draw from these cases? First, we have seen that failure is often a by-product of risky ventures. Projects often involve untested or novel technologies and processes. Risk of technical failure is always present in these circumstances. Further, projects upset the status quo of the organization. They operate outside formal channels with temporary groups of diverse individuals pulled together for one purpose. They often violate political relationships and established chains of command. Given the environment within which many projects operate, it is not surprising that failure is a very real possibility with any project undertaking.

The second conclusion is that past failure need not discourage us from future efforts. Indeed, it is through these past failures that we gain the experience and wisdom to push on toward successful conclusions. There are two equally erroneous responses managers can have toward past failure. The first is to brush it aside with as little thought as possible; in effect, push it out of sight and out of mind. The other error is the mirror opposite: to become so focused on past failure that it prevents an organization from taking the necessary steps for new ventures and project start-ups. Consequently, these firms suffer from a form of paralysis that precludes quick responses to competitors, to say nothing of their inability to move proactively. It is important not to become a victim of past failure, either through a mulish unwillingness to learn from it or through excessive timidity in trying again.

Our hope is that this book will offer the middle ground: Charting some examples of famous (and not so famous) project failures, primarily in order to point to the relevant lessons that can be derived from their study. We believe that no project is worth analyzing if there is no object lesson that it can offer. This book, *What Made Gertie Gallop*, is a collection of object lessons, of many different types and many different projects. They all share one common characteristic—through some degree of management error, they were failures. This last point is important: There are lessons to be learned from failure, if only we are willing to draw them. The first step is to learn precisely what project "failure" has come to mean. That is the easy part. Far more difficult is taking the next logical step and looking inside for the causes, particularly when those errors bear an uncomfortable similarity to our own past experiences. It is, however, in this honest assessment, that we can come closest to deriving the power from this book and its lessons.

References

Milbank, D. and Kamm, T. 1995. *Wall Street Journal*, "Eurotunnel suspends interest payments," vol. 76 (234), September 15, p. A7.

Brockner, J. 1992. "The escalation of commitment to a failing course of action: Toward theoretical progress," Academy of Management *Review*, 17, pp. 39–61.

Cleland, D.I. 1988. "Project stakeholder management," in D.I. Cleland and W.R. King (eds.), *Project Management Handbook, 2nd ed.*, New York: Van Nostrand Reinhold, p. 275–301.

Evans, J.R. 1993. *Applied Production and Operations Management, 4th Ed.* St. Paul, MN: West Publishing Co.

Obermeyer, N.J. 1990. *Bureaucrats, clients, and geography: The Bailly nuclear power plant in northern Indiana*, (Research paper no. 216). Chicago, IL: University of Chicago, Department of Geography.

Index